Model Risk Management with SAS®

SAS Institute Inc.

§sas.

sas.com/books

Contents

About This Book

What Does This Book Cover?

Model risk can occur when a model has inappropriate data, specifications, assumptions, calibration, or other errors. Proper model risk management can mitigate this risk through the development, validation, and governance of models across the entire organization. SAS Model Risk Management is a web-based solution that helps your organization capture and manage information that you can use for model inventory management, model validation, ongoing monitoring and maintenance, change management, and model usage documentation.

The training in this book prepares members of your organization's project team to be effective and informed participants in the requirements development and solution design phases of your implementation. The book includes hands-on demonstrations, and it teaches key concepts, terminology, and base functionality that are integral to SAS Model Risk Management.

The first chapter of this book provides an overview of the regulatory environment, and it introduces the SAS Model Risk Management platform. This chapter is based on videos narrated by David Asermely and Peter Plochan, which are available on the SAS video site at https://video.sas.com/category/videos/sas-model-risk-management. The remainder of the chapters are a step-by-step introduction to using SAS Model Risk Management based on the course SAS® Model Risk Management: Workshop from SAS Education.

In addition, this book includes an Appendix that provides an overview of SAS Model Implementation Platform (MIP). SAS MIP executes complex and computationally intensive modeling systems for bank stress tests and for other enterprise-level risk assessments. SAS MIP is a web-based solution that helps you streamline the model estimation-to-implementation process, dramatically reduce model run times, and improve auditability.

Is This Book for You?

SAS is the largest and most successful model risk management software vendor in the world. Software like SAS Model Risk Management and SAS Model Implementation Platform enable organizations to use a data-driven approach to assess their model risks, identify gaps, review and update their policies and procedures, and reassess their risks. This iterative approach continually improves their model quality and efficiency. If you want to learn more about the model risk management life cycle, as well as fine tune your governance policy and approach, then this book is for you.

What Are the Prerequisites for This Book?

Familiarity with SAS Model Risk Management is helpful to understand this book. Access to the software is necessary to follow along with the demonstrations.

What Should You Know about the Examples?

This book includes tutorials for you to follow to gain hands-on experience with SAS.

Software Used to Develop the Book's Content

SAS Model Risk Management 7.4, SAS Model Implementation Platform 3.2, and SAS Visual Analytics were used to develop the content and examples for this book.

We Want to Hear from You

SAS Press books are written *by* SAS Users *for* SAS Users. We welcome your participation in their development and your feedback on SAS Press books that you are using. Please visit sas.com/books to do the following:

- Sign up to review a book
- Recommend a topic
- Request information on how to become a SAS Press author
- Provide feedback on a book

Do you have questions about a SAS Press book that you are reading? Contact the author through saspress@sas.com or https://support.sas.com/author_feedback.

SAS has many resources to help you find answers and expand your knowledge. If you need additional help, see our list of resources: sas.com/books.

Chapter 1: Introduction to SAS Model Risk Management

Overview

In this chapter, you will be introduced to SAS Model Risk Management. Model governance and model risk management is getting increasingly demanding. The number of models is rising based on both regulatory and business demands. The amount of input data is growing and getting more complex. And on top of that, the authorities are placing stricter demands on the organization's ability to keep the overview and make fact-based decisions around the governance of models and the mitigation of model risk.

SAS Model Risk Management enables all stakeholders in the model life cycle – developers, validators, internal audit, and management – to get overview reports as well as detailed information in one central place. Not only does the solution ensure that the model validation team can easily provide quick replies to questions and inspections from the authorities, but it also provides business benefits through better overview and higher efficiency in the model life cycle.

With SAS Model Risk Management, there are two key focuses that drive our product. The first is delivering efficiencies and other benefits to modeling teams. We know that banks spend a tremendous amount of resources on their models. Having a solution that helps organizations work more efficiently is one of the key business drivers behind the solution. The other, of course, is reducing the risk associated with those models.

SAS Model Risk Management goes above and beyond legacy and operational risk approaches. It is driven by robust qualitative and quantitative components. Combined with the integration with the entire SAS modeling infrastructure, the product allows all of those components to work in a much more efficient way.

Regulatory Environment

When we look at what is currently happening in the banking industry, we can clearly see that banks are being challenged on multiple fronts simultaneously. First, there is an increasing demand for more model governance and model risk awareness. This is coming from both the business side, pushing for high utilization of models in banks' decision-making, and also from the regulatory side where the recent guidelines have put banks under unprecedented scrutiny. At the same time, banks are pushed to faster model deployment and increased model performance. Lastly, in peril, banks are required to build new models. A very good example of this is the IFRS 9 impairments.

According to some senior model risk managers, IFRS 9 will have a significant impact on banks' MRM processes. Compared to the current impairment regime, IFRS 9 will have more financial impact and will require more impairment provisions, which will have to be recalculated more often to capture the changes in credit quality. The impairments calculation will be more complex, will require the collection of more data, and will require the development of more models. IFRS 9 principles offer more alternative interpretation that will define the new methodology behind these new models. All of this work will require much more internal cooperation between risk and finance. And everything will have to be accepted by the statutory auditor. Therefore, robust governance around all this new data – models, interactions, and calculations – will be crucial. The impact of IFRS 9 should not be underestimated when thinking about the future of banks, model risk management, and model risk governance processes.

IFRS 9 is just one piece of the puzzle in all these developments, though. The SAS Model Risk Management solution is designed to address not only the current challenges that banks are facing now, but also the ones that are yet to come.

Design Principles

Breaking down the design principles of SAS Model Risk Management, there are tools within the solution that allow the collection of information in a very straightforward and easy way that builds a robust inventory of all of the models within an organization. There are then tools that help conduct the proper assessments, validations, of those models in a way that is straightforward, easy-to-follow, and intuitive. The solution then takes that information and provides it to the user in succinct reports the organization can use to understand where it has model risk and then fine-tune model risk management practices in a way that produces optimum results.

Table 1.1: SAS Model Risk Management Design Principles

Collect	Inventory	Conduct	Report
Capture data via easy-to-use web interface	Store metadata of model universe: models, validations, non-models, data sources, findings, assessments, etc.	Conduct end-to-end model life cycle via workflows	Interactive dashboards for executive management

Collect	Inventory	Conduct	Report
Easy onboarding of legacy data	Classify models using tags	Set up tasks for three lines of defense	Drill-down into model risk drivers
Make bulk changes to data	Store documents and artifacts for all model life stages	Automated alert/notification dispatcher (real-time/timed)	Aggregate model risk across the firm
Setup integration with external systems	Assign stakeholders	Track and audit changes	Set up KRIs and EWS dashboards
Easily collect model monitoring metrics	Set up 360 links for model	Security based on entitlement	Full integration with Microsoft Office Applications
		Take bulk actions (sign-offs)	

Within SAS Model Risk Management, there are a number of components. These components all come together to tell the true model risk story.

Figure 1.1: MRM Architecture

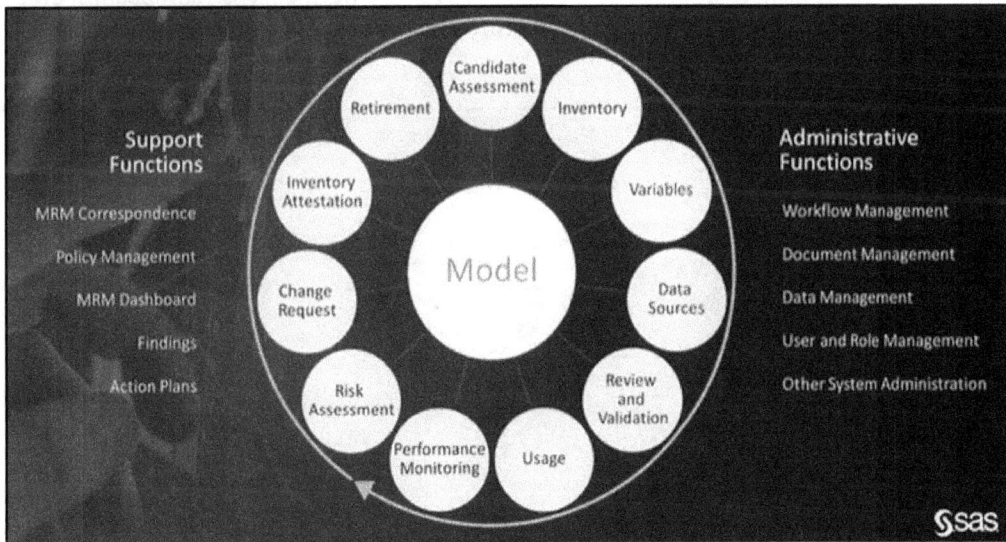

In a nutshell, the SAS solution is a robust model information system that tracks all the relevant information and stakeholder involvement across the whole end-to-end modeling life cycle. It establishes linkages between the relevant components of the modeling universe, for example, between a model feeding another model or between models and their usages. All interactions between various stakeholders and approval cycles follow the underlying workflow to define which information has to be provided or which action has to be taken by whom and when. Once all this information is collected and stored in a secured fashion, various reports can be easily built using the intuitive self-service reporting engine.

All of the reports, workflows, and screens or data definitions are fully configurable by the business user without the need for IT involvement or any hardcoding. Out-of-the-box content enables users to start quickly with content based on the industry best practices that can evolve as users learn what best fits their needs. By having such information systems in place, users can quickly identify areas with concentration of model risk and focus their attention there. By following predefined workflows, processes are streamlined, manual handovers are reduced, and efficiencies are achieved by sharing information among the relevant stakeholders. Last but not least, users should be ready for any questions about the models that might come from external or internal stakeholders.

Example

Let's look at a real-life example of SAS Model Risk Management in action. Assume that you are a model governance officer who wants to manage and monitor model risk. First, you would need to know where the model risk resides in your organization and what is driving it.

Model Risk Dashboard

For the purpose of managing and monitoring risk, you can use the model risk dashboard shown in Figure 1.2, which you can find in the **Model Risk Summary** tab on the dashboard.

Figure 1.2: Model Risk Dashboard

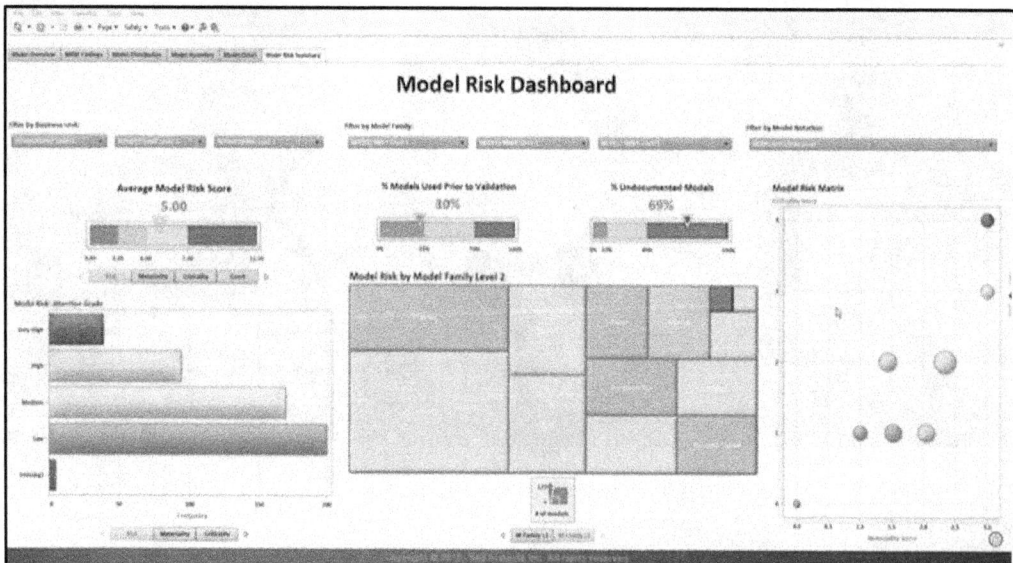

This dashboard uses a model risk quantification framework where every model is scored on importance or materiality (in other words, quality or criticality). By combing these two scores, each model can be plotted on a Model Risk Matrix, which enables you to quickly spot not only the models that are most important for your organization (in other words, having the highest materiality), but also the ones that require immediate attention (criticality).

If you click on the red circle in the upper right quadrant of the Model Risk Matrix, you can see that there are 39 models that have both highest materiality and highest criticality scores. (See Figure 1.3). This is where we need to focus.

Figure 1.3: Model Risk Matrix

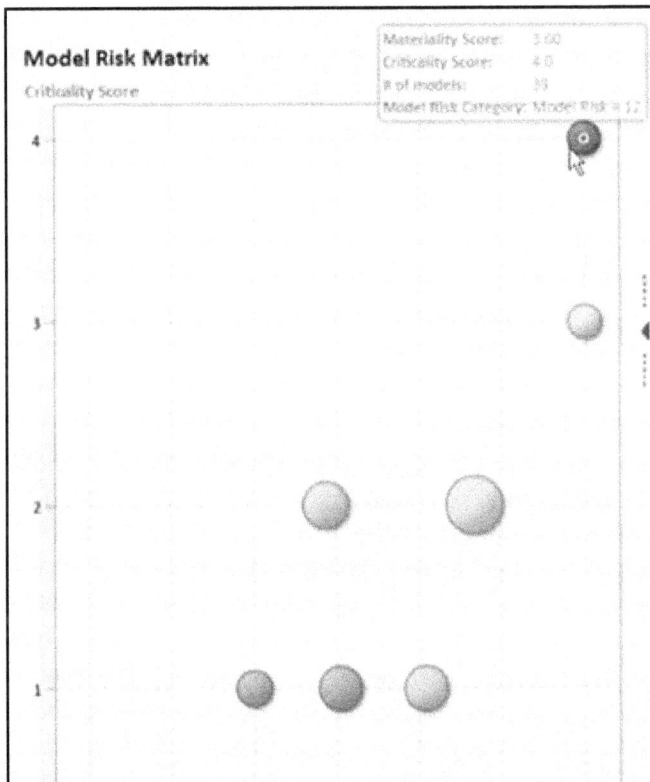

Furthermore, by aggregating the two scores, first on an individual level, and then across relevant models of portfolios, you can monitor model risk concentrations. For example, in the % Undocumented models section of the dashboard, if you click on the red square in the upper right corner, you can clearly see that the derivatives model portfolio requires immediate attention. (See Figure 1.4.)

Figure 1.4: % Undocumented Models

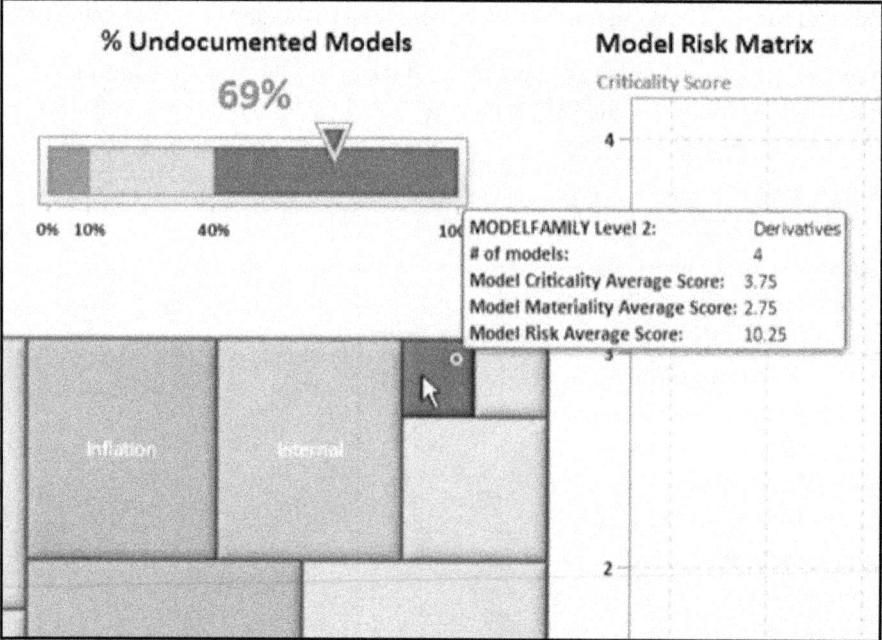

You can easily zoom in to a particular business unit and explore the model of risk concentration there. In the upper left of the dashboard, click **Filter by Business Unit** to select a business unit from the drop-down menu. In this example, we can see that within the Sunshine Financial Business Unit, it is the inflation model portfolio that requires attention. (See Figure 1.5.)

Figure 1.5: Dashboard Filtered by Business Unit

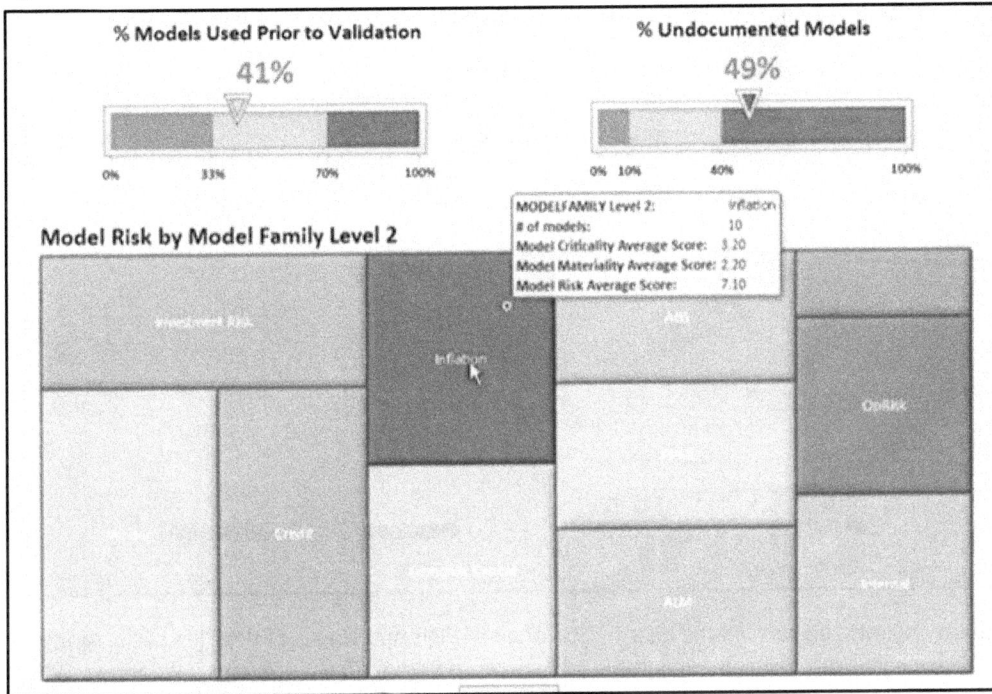

Model Governance Overview Dashboard

Besides the model risk quantification information, there is other useful information tracked in the system to support governance processes. Let's now assume that you are the manager of the model validation team. You want to plan model validations for the next six months. In order to do that, you want to get an idea of how many models are currently being developed because your team will need to validate them. For this, you will use the Model Governance Overview section of the dashboard, which is the leftmost tab on the dashboard. (See Figure 1.6.)

Figure 1.6: Model Governance Overview Dashboard

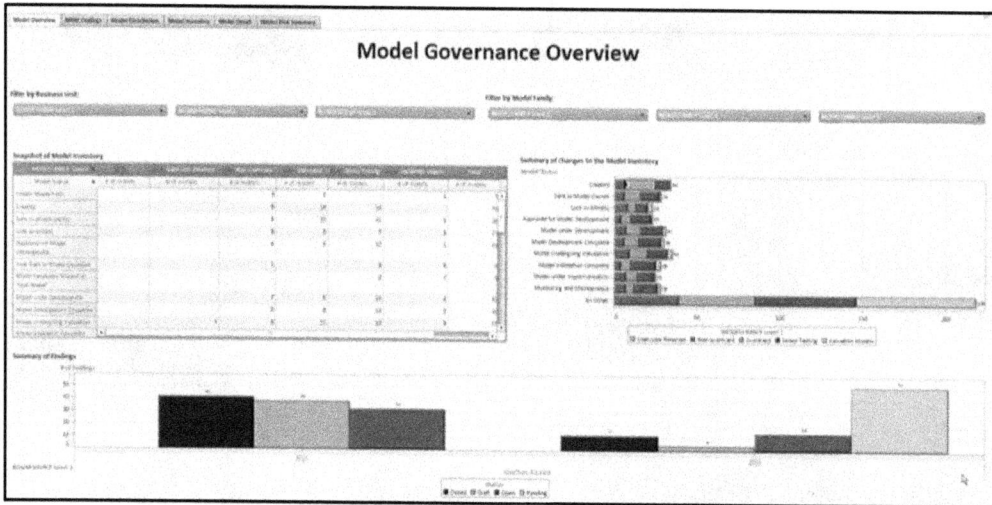

In this overview, you can see the distribution of all models and stages of their life cycle. Along the bottom, you can also see the number of registered findings from previous validations. What might concern you is the unusually high number of pending findings that have not yet been accepted by the owners.

For now, let's focus on the new models that are going to be developed. In the **Summary of changes to the Model Inventory** section, we can see that there are 34 new models. (See Figure 1.7.)

Figure 1.7: Summary of Changes to the Model Inventory

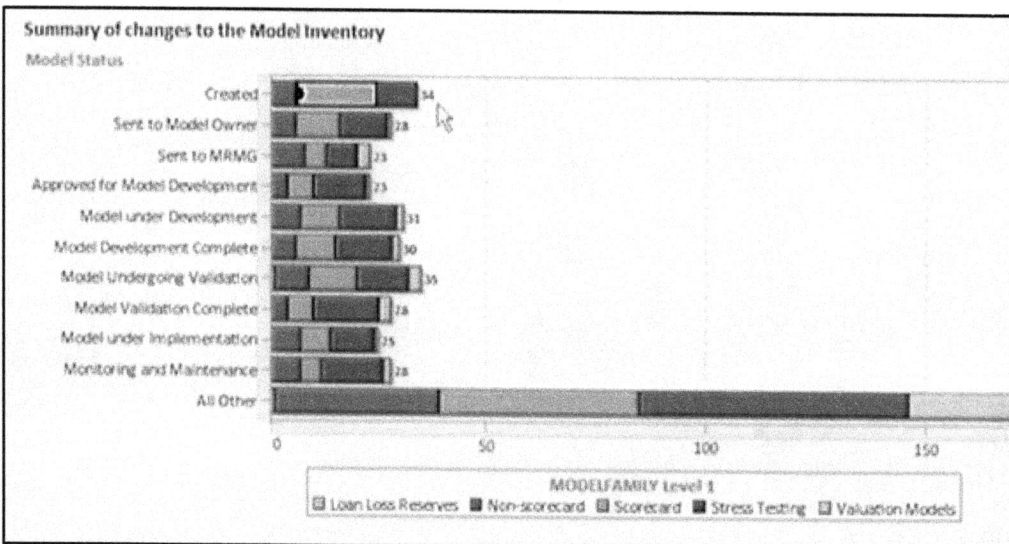

Let's say that we are particularly interested in scorecard models because they require the most effort to get validated. If you click on the middle bar of Created, we can see that there are 18 of these models. Right-click on the bar and select **Drill Down in Scorecard** from the pop-up menu. This enables you to zoom into the model inventory. (See Figure 1.8.)

Figure 1.8: Model Inventory

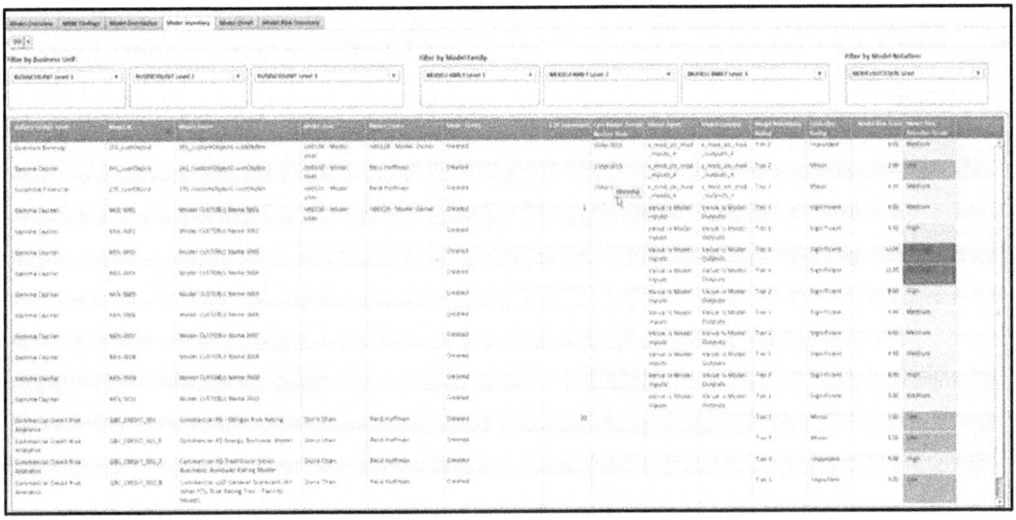

In the Model Inventory section, you can see the selected scorecard models alongside key information such as model risk score. We can also see that one model has 20 associated sub-models, but it is still marked as low risk. But what if, for example, you believe that a model with that many linked sub-models should get a higher criticality rating? To change the rating, click on the row of the model that you are interested in. Then, go to the **Model Detail** tab across the top of the screen. This tab enables you to see more information about a particular model as shown in Figure 1.9.

Figure 1.9: Model Detail

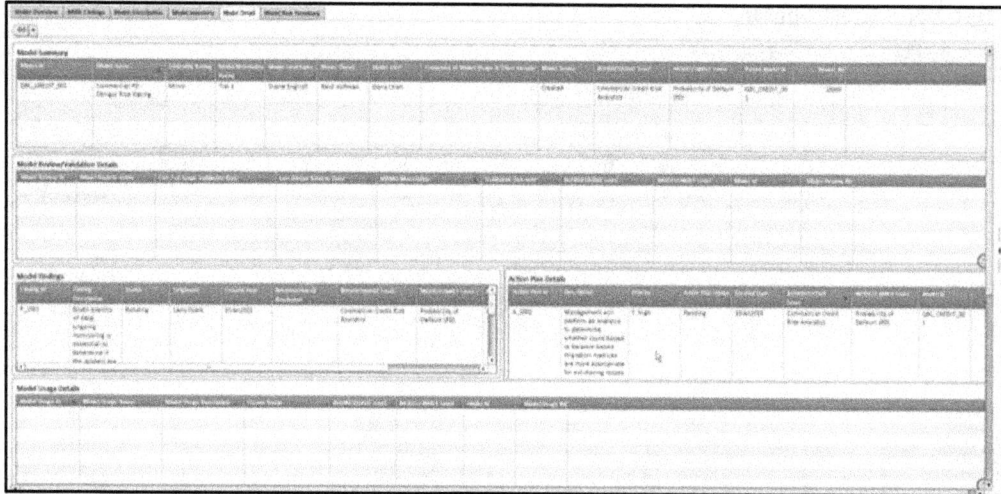

Some of the sections in this tab include associated findings, action plans, and so on. Click on the model in the **Model Summary** section at the top. This takes you into the operational environment where the model information is managed. You can edit certain model properties such as rating in this environment, as explained in the next section.

Operational Environment

Now, let's have a quick look at the environment where you can view, review, and edit the information about selected models. Your capabilities depend on your permissions and the stage of the model life cycle. In Figure 1.10, you can see that the selected model is currently in the Created stage. You can also see a description of the type of model and the model itself. By scrolling down, you can see the key stakeholders linked with this model.

Figure 1.10: Operational Environment

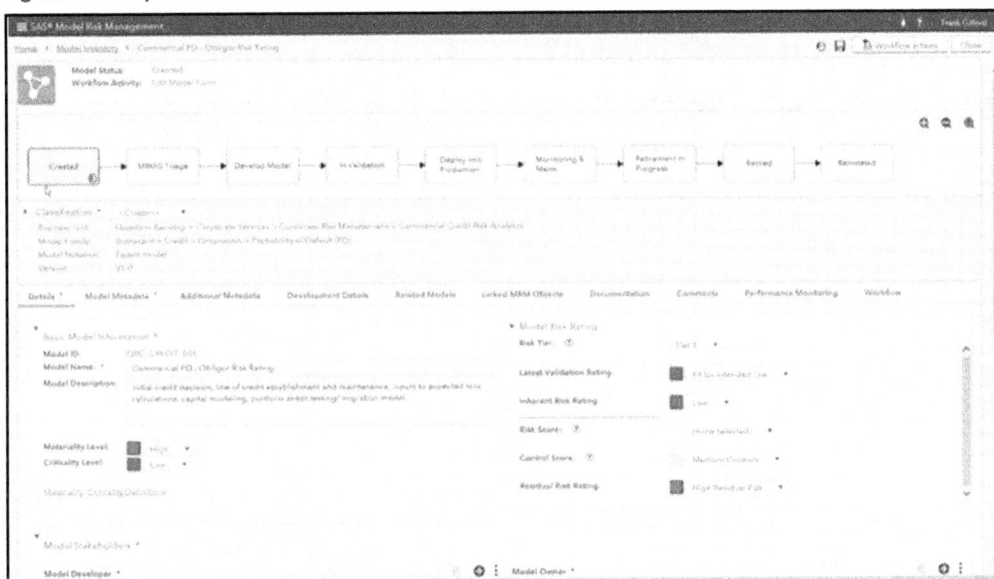

Let's look at the rating information. In this example, we want to change the criticality level from Low to Moderate. Click on the drop-down menu next to **Criticality Level** in the **Basic Model Information** section. Select **Moderate** and then click out of the menu.

The Operational Environment allows you to store and track various information about a model such as linked models. To view models with a relationship to the selected model, click the **Related Models** tab. In this tab, you can see the 20 sub-models that are associated with this model. (See Figure 1.11.)

Figure 1.11: Related Models

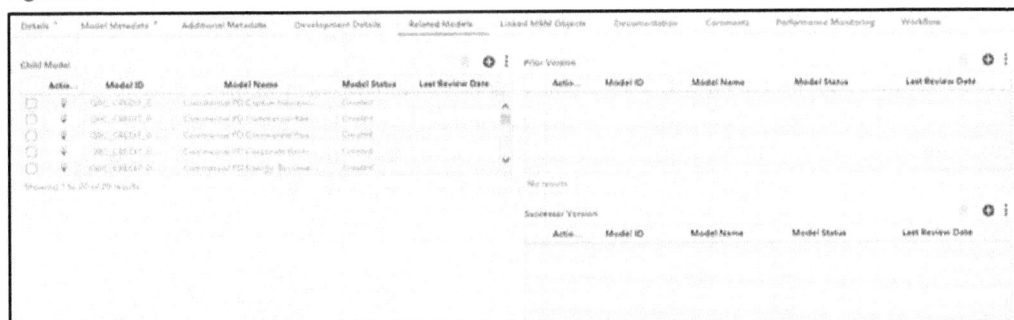

You can also link your model with other objects such as findings and models reviews. In the **Documentation** tab, you can store model documentation or model performance reports.

Depending on your role and stage of the model's life cycle, the underlying workflow defines your available actions. You can view this information by clicking the **Workflow** tab. (See Figure 1.12.)

Figure 1.12: Workflow

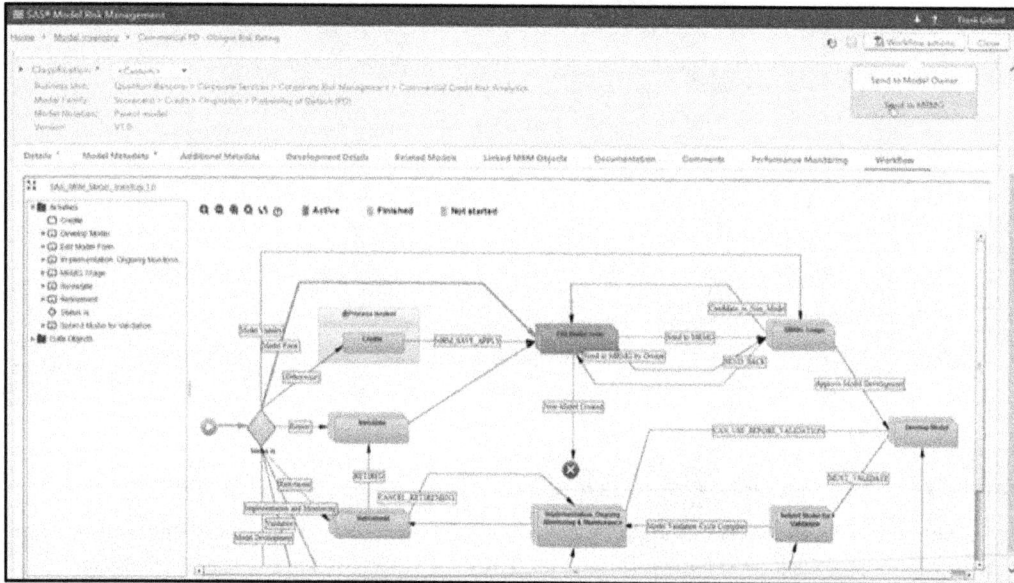

You will receive notifications when you are expected to perform a particular task. At any time, you can see where you are within the workflow. In the upper right corner of the screen, you can see what your potential actions are by clicking the **Workflow actions** button. At this point, your options are to send the model form to the model risk management team or send it to the model owner.

Summary

Everything that you have seen with SAS Model Risk Management – workflows, reports, page definitions, and metadata definitions – is fully configurable without the need to do any coding. This solution can help your institution decrease model risk and reduce your workload while increasing efficiencies and awareness and compliance with regulations.

Chapter 2: Model Life Cycle

Overview

The term *model* refers to a quantitative method, system, or approach that applies statistical, economic, financial, or mathematical theories, techniques, and assumptions to process input data into quantitative estimates.

A model consists of three components:

- an *information input* component, which delivers assumptions and data to the model

- a *processing* component, which transforms inputs into estimates

- a *reporting* component, which translates the estimates into useful business information

Model risk can result in financial and reputational loss. Adverse consequences from misused model output have various causes, which can include flawed assumptions, bad design, erroneous inputs, incorrect implementation, and inappropriate usage. Complex models for new products and markets make this an ever-bigger challenge to manage. Non-compliance can result in regulatory fines, increased capital charges, and headline risk. Model risk cannot be eliminated, but it can be managed with a proper control environment and governance structure.

A model life cycle needs to be managed by a well-defined framework and system. You can manage model risk like any other type of risk with sustainable systematic processes by adhering to the following core processes (see Figure 2.1):

- Identify, estimate, monitor, and manage model risk
- Robust internal controls and effective governance
- Set up and maintain model inventory
- Integrate model risk limits with risk appetite
- Comprehensive and sustainable model risk management program
- Robust, efficient, auditable applications environment

Figure 2.1: Model Life Cycle

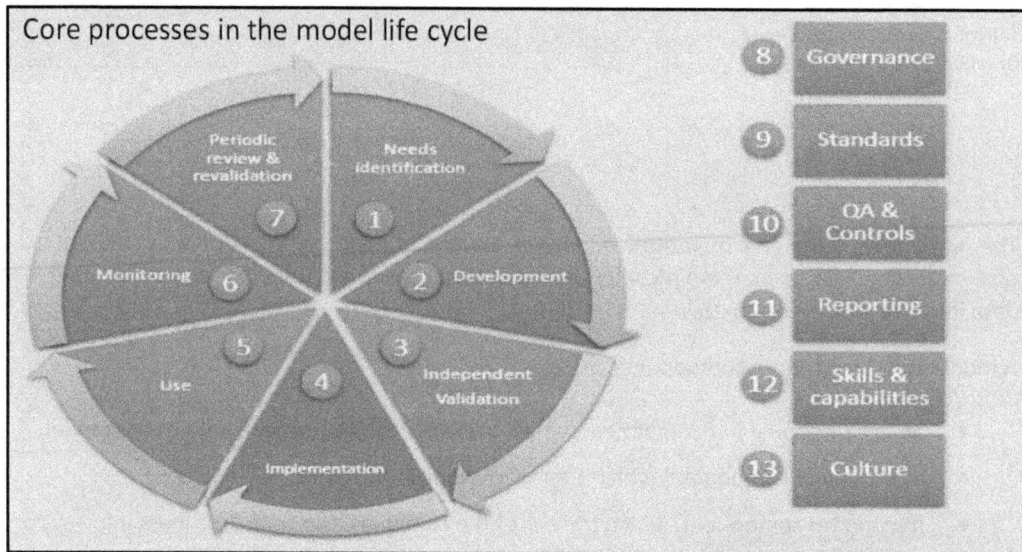

No model is an island. Model stakeholders can include the owner, developer, validator, model risk management oversight, senior managers, internal audit, and regulator.

In addition, models are associated with a lot of other information. Related information can include the following:

- Related models
- Related findings
- Related data and variables
- Related action plans
- Related documents
- Related usages

- Related model reviews
- Related policy exceptions
- Related correspondence

Keeping track can be tricky, so a good system is needed! This is where SAS Model Risk Management comes in.

SAS Model Risk Management

According to the IIA, there are three lines of defense that encompass the following functions:

1. Functions that own and manage risks.
2. Functions that oversee risks.
3. Functions that provide independent assurance.

Each line of defense has a distinct role within the organization's wider governance framework.

SAS Model Risk Management supports the full model life cycle in a single solution. As shown in Figure 2.2, it supports the needs of all lines of defense – including an integrated dashboard application that allows senior management to access all aspects of every model in the inventory including inter-model relationships.

Figure 2.2: Three Lines of Risk Management Defense

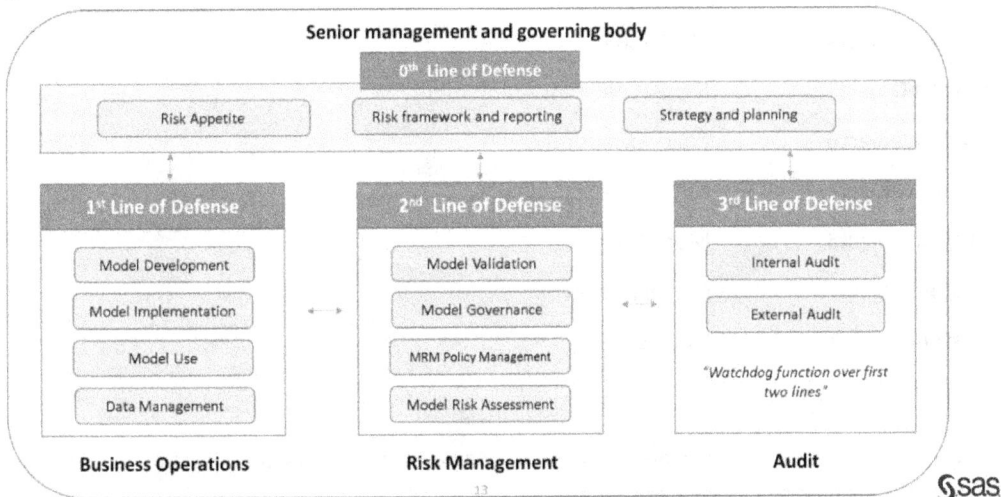

SAS Model Risk Management runs on SAS 9.4. It is a user-friendly, web-based application that facilitates the capture and life cycle management of statistical model-related information. That information is then used to conduct all aspects of model risk management, including governance. Specifically, SAS Model Risk Management facilitates the entry, collection, transfer, storage, tracking, and reporting of models that are drawn from multiple lines of business across an organization. It also integrates with other SAS products, including the SAS Workflow Engine and SAS Visual Analytics.

SAS Model Risk Management has three key components: the user interface, workflow, and dashboards. Over the next few chapters, we take an in-depth look at each of these components.

Key Concepts

Objects

SAS Model Risk Management 7.4 is built entirely from custom objects. Behind every menu item and window in Model Risk Management sits a business object. Examples of business objects are models, model reviews, and findings. In the back end (that is, in the configuration areas such as SAS Management Console and the database), each business object is related to a custom object. Model is CustomObject1, ModelVariable is CustomObject8, and so on, as shown in Table 2.1.

Table 2.1: Business Object Examples

Custom Object	Model Risk Management
CustomObject1	Model
CustomObject2	ModelUsage
CustomObject3	Non-Model
CustomObject4	ModelChangeManagement
CustomObject5	ModelCandidateAssessment
CustomObject6	PolicyException
CustomObject7	MRMCorrespondence
CustomObject8	ModelVariable
CustomObject11	Model Review
CustomObject12	Finding
… and more	

Only admin users need to be aware of the underlying custom objects, and users need to know only the business objects. (See Figure 2.3.)

Figure 2.3: Business Objects

Business Object

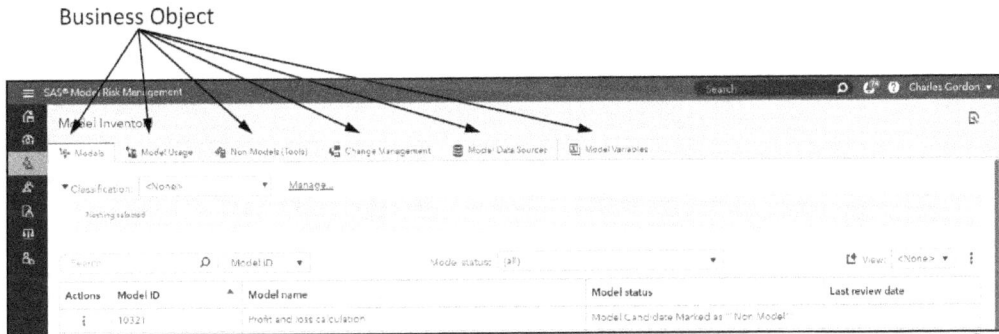

Links

In Model Risk Management, one or more link types can be created between any two business objects. This concept is known as *360-degree linking*. Linked objects appear in tables on the main object screen.

Dimensions

Organizations have a number of different layers and hierarchies under which they operate. In SAS Model Risk Management, these hierarchies are referred to as *dimensions*.

Dimensions enable you to perform the following tasks:

- categorize and filter business objects
- fine-tune reports
- control and constrain the tasks users can perform (by assigning users to a role and a dimensional scope)

The solution ships with out-of-the-box dimensions (15) but can create up to 23 unique dimensions.

> **Note:** Most organizations normally need between four and eight dimensions only!

Dimensions specify the location and operational areas of your data in Model Risk Management. Positions assigned to users determine their scope. (We will look in more detail at users in the next section.) Dimensional location of data and user scope determined by assigned positions determine whether a user can see data. Some examples of dimensions include Geography, Model Family, or Product. For Geography, a model could be assigned a dimensional location of North America. Users with a position outside of North America would not be able to see or work with the North American models.

Users, Roles, and Positions

When a user logs on to SAS Model Risk Management, the application examines user positions in the database. Each position is defined by a scope and a role. Each role has capabilities that are associated with it. Therefore, the *role* determines what you can do in the application. The *scope* determines where you can do things in the application.

One user can be assigned one or more positions. Positions apply at and below the defined dimensional node. By assigning a position at BU 1 level, a user is able to perform the tasks defined by the Model Owner role at the BU 1, Department 1, and Department 2 nodes. For a user to perform any task, that user must be assigned at least one position. You can batch load positions for multiple users by using the Positions data loader.

Figure 2.4 illustrates the relationship between user, role, and position.

Figure 2.4: User, Role, and Position

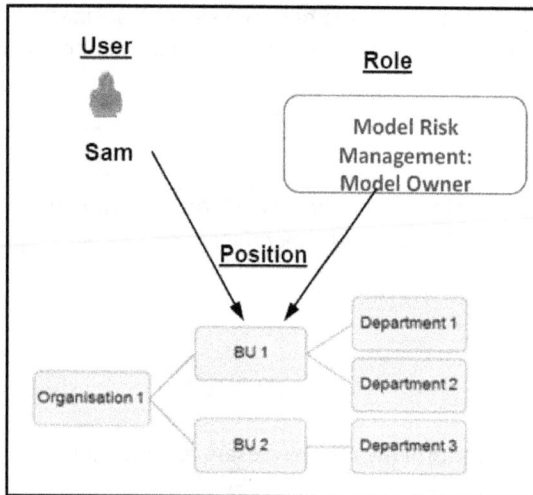

Data Loader

Data in Model Risk Management is typically entered by users via the User Interface. However, administrators can load data using the data loader functionality. This is valuable for situations where data is most efficiently loaded in bulk.

There are two types of data in Model Risk Management, reference and transactional, as shown in Table 2.2.

Table 2.2: Types of Data

	Reference Data	Transactional data
Characteristics	Not updated frequently Typically managed by central group Required before users can work with the system	Updated frequently Typically managed by the end users
Examples	Dimensions Users Roles Positions Exchange Rates	Models Model Reviews Findings Action Plans Model Risk Assessments

In the loader Excel spreadsheet, columns are fields (or other system items) and each row is an entry. The loader checks typical database restrictions (duplicate IDs, data types, and so on). If the loader includes an instance that already exists in the solution, this results in an update. Data can also be unloaded similar to loading.

Databases have rows, columns, and tables. Data loaders enable you to alter those from the front end.

Practice: User Interface Tour

1. View the home page.
    a. Sign in as **chunter@saspw** using the password **Orion123**.
       The first things you see are the tasks and notifications for this user.
       Notice that Carrie has five notifications, and below that is a list of her models.
    b. How many total models does she own?
       Answer: _____
    c. Scroll down to see her model reviews, model findings, and custom links.

2. View the Model Inventory page.

   a. Click the **Model Inventory** icon.

   b. Click the **Models** tab.

      Notice that there are six tabs.

      1) How many models are in the system?

         Answer: _____

      2) How many are showing?

         Answer: _____

      3) Clicking one of these will take you directly to the model page for that model.

   c. Click the **Model Usage** tab.

      Notice that there are six different model usage cases here.

3. View the Model Reviews page.

   a. Click to access the Model Reviews page.

      Notice that there are 10 model reviews.

   b. Who is the last editor on all of them?

      Answer: _____

4. View the Model Monitoring page.

   Click the Model Monitoring tab.

   Notice that there are two examples shown here.

5. View the Findings page.

   a. Click to reach the Findings page.

   b. How many total findings are there?

      Answer: _____

      When there are findings, there are usually also action plans.

   c. Click the **Action Plans** tab.

      1) Notice that there are 30 total action plans.

      2) Scroll through them all, and note that none of them are closed.

6. View the MRMG Assessments page.

   a. Click to reach the MRMG Assessments page.

      Notice that the first tab is Model Risk Assessments.

   b. How many of these have been performed?

      Answer: _____

   c. Also, notice that the model inventory attestation, candidate assessments, and policy exceptions are here.

7. Sign out.

8. Close Chrome.

Practice: Solutions

How many total models does she own?
Answer: 11

How many models are in the system?
Answer: 287

How many are showing?
Answer: 18

Who is the last editor on all of them?
Answer: Frank

How many total findings are there?
Answer: 52

How many of these have been performed?
Answer: 41

Chapter 3: Model Onboarding

Model Candidate Assessment

Before a model development candidate can be entered into the model inventory, typically the candidate must be assessed by the model risk management group (MRMG) to ensure that it meets organizational or regulatory criteria for a model.

The best practices outlined below were informed by implementing SAS Model Risk Management and interactions with practitioners at various US banks. Business managers in a bank use different computational tools to support their business operations and decision making. However, not all tools are models, such as a spreadsheet that estimates an output using a closed-form formula. In order to give a model the right attention and to ensure it meets organizational and regulatory criteria for a model, the tool/candidate needs to pass a "What's a model?" test.

The SR 11-7 guidance offers a set of five tests that candidates must pass.[1] Many banks augment these tests with bank-specific criteria to qualify a candidate as a model. In general, all tests must be passed before a candidate is qualified as a model or not. A non-model is sometimes referred to as a user-defined tool (UDT) or end-user computing (EUC). SAS Model Risk Management provides a receptacle to manage an inventory of these non-models. In cases where one or more non-models provide a service to a managed analytical model, that relationship is memorialized via a 360-link.

Model Assessment

SAS Model Risk Management provides a questionnaire incorporating the above model-definition criteria typically completed by the second line of defense (the MRMG). The MRMG uses the model candidate information provided by the model owner to answer the questions. The candidate assessment process helps in tracking how the model candidate assessment was conducted and why the model candidate met or did not meet the model criteria. The assessment process therefore facilitates accountability and creates a traceable information stream for auditing and reporting purposes. The inventory process for a model is shown in Figure 3.1.

Figure 3.1: Model Inventory Process

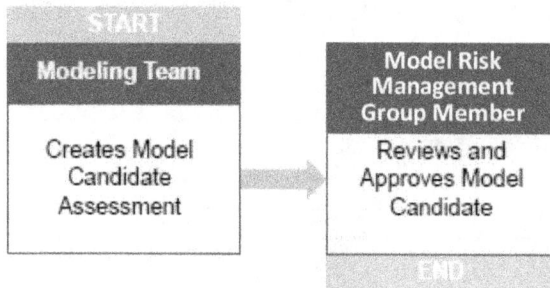

Demo 3.1: Viewing the Model Candidate Assessment

1. Assess a model candidate.

   a. Start Chrome.

   b. Sign in as Carrie Hunter (**chunter@saspw** and the password **Orion123**).

   c. Click the **MRMG Assessments** category button.

   d. Click the **Candidate Assessment** page.

   In the Candidate Assessment table, click **New** to create a new candidate assessment.

   e. Expand the **Classification** selector.

   f. Click **Manage**.

   1) Under Business Units, select **Sunshine Financial**.

   2) Under Model Families, select **Loan Loss Reserves** ▶Credit.

   3) Click OK.

   g. In the **Model Candidate Assessment Name** field, enter **Sunshine Credit Loss Reserves.**

   h. In the **Model Candidate Assessment Description** field, enter **Monitor the loss reserves for credit accounts**.

   i. Enter **4/1/2019**, or something more appropriate, for **Assessment Due date**.

   j. Click the **Model Criteria** tab.

   1) Under Model definition criteria, for question 1, click **Yes**. For **Justification**, enter **It is Needed**.

   2) For Question 2, click **Yes**. For **Justification**, enter **It has a financial impact**.

   3) For Question 3, click **Yes**. For **Justification**, enter **There are some decisions made on judgment**.

      4) For question 4, select **Yes**. For **Justification**, enter **Approximates reality as near as possible**.

      5) For question 5, select **Yes**. For **Justification**, enter uses **Bank data in the calculations**.

k. Click the **Save** icon.

l. From the **Workflow actions** menu, select **Send for Approval** to send the model candidate assessment to the MRMG for approval.

m. Sign out as Carrie Hunter.

n. Sign in as **kfeldman@saspw** (Kirsten Feldman) with the password **Orion123**.

o. Model Risk Management should be selected. Click **Continue**.

p. Under My tasks and notifications, find the model candidate assessment that you just created.

q. Click on that candidate assessment.

      1) Kirsten has found no problems. Click **Workflow actions**.

      2) Select **Approve**.

r. Sign out of Model Risk Management.

End of Demonstration

Non-Model (Tools) Assessment

Tools that are developed must meet certain criteria, as determined by regulatory or organizational standards, in order to be categorized as models. Other tools might not meet some of the criteria of a model but are still considered useful and necessary to the organization. This applies to business rules, key model assumptions and limitations, or simple calculations. For example, a spreadsheet that details profit and loss information, a useful calculation to the organization, does not apply any level of uncertainty or judgment. Therefore, it is not considered a model. SAS Model Risk Management uses the term *non-model* to describe these items.

Figure 3.2 shows the non-model inventory process.

Figure 3.2: Non-Model Inventory Process

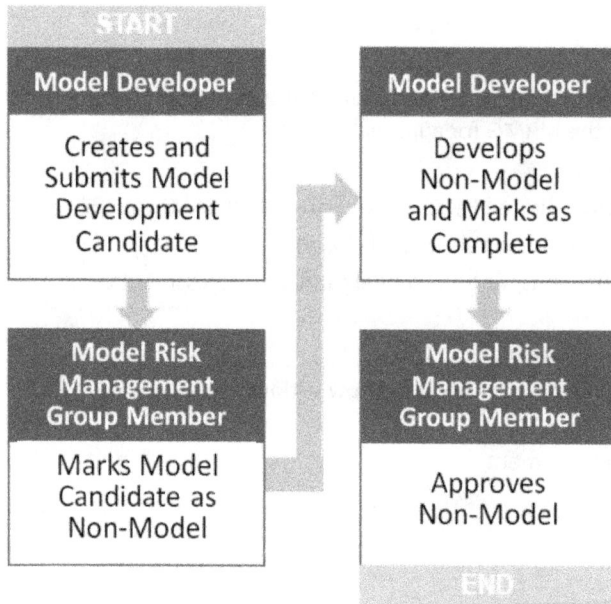

Demo 3.2: Viewing the Non-Model (Tools)

1. Use the non-model inventory process.
   a. Sign in as **Chunter@saspw** using the password **Orion123**.
   b. Click the **Model Inventory** category button from the navigation bar.
   c. On the **Model Subcategory** tab, click **New** to create a new model.
   d. For **Classification**, click **Manage**.
      1) Expand **Model Families** to **Valuation Models ▶ Inflation**. The business unit should already be **Sunshine Financial**.
      2) Click **OK**.
   e. Enter **Inflation Valuation Model** as the model name.
   f. Leave the description blank.
   g. Scroll down to **Model Stakeholders**. The required fields should be populated.
      1) Model Developer – **Carrie Hunter**
      2) Model Owner – **Carrie Hunter**
      3) CoE coordinator – **Paul Melhoff**
      4) Model User – **Ann Kellog**
      5) MRMG Group Member – **Kirsten Feldman**
      6) OpRisk Manager – **Raul Nunzio**

    h. Click the model's **Metadata** tab.

        1) Carrie believes this is a model. Select **No** for **Is this a Non-Model?**.

        2) Select **Bank Developed** for the **Vended Type** field.

        3) Select **No** for the **Model to be used before validation?** field.

        4) Selects **MS-Excel** for the **Model Development Software/Tools Used** field.

    i. Click **Save**.

    j. From the Workflow actions menu, select **Send to MRMG** to send the model to the MRMG for approval.

    k. Sign out of Model Risk Management.

    l. Sign in as **kfeldman@saspw** using the password **Orion123**.

    m. From the Task list on her home page, select the **Determine Model Viability for the Inflation Valuation** model.

    n. Click the model Metadata tab.

        1) For **Is this a Non-Model?,** change the answer to **Yes**.

        2) Enter **This spreadsheet does not contain any assumptions and does not use statistical methods. It is therefore not a model.** for the explanation.

        3) Enter **March 1, 2019** as the date entered in inventory.

    o. From the **Workflow actions** menu, selects **Mark as Non-model**.

    p. Click **Save** and enter a reason in the **Change Reason** field.

    q. Sign out of Model Risk Management.

2. Develop the non-model.

    a. Sign in as **chunter@saspw** using the password **Orion123**.

    b. On her home page, in the Task list, find and select **Create and link non-model form for the Inflation Valuation Model.**

    c. Click the **Linked Objects** tab.

    d. In the Non-Model table, click **New** to create a model.

        Leave the classification as it is.

    e. On the Details tab, enter the following:

        1) Enter **Model for Inflation Valuation (non-model)** for the non-model name.

        2) Leave the description blank.

    f. Click **Save** to save the non-model.

    g. Click **Workflow actions** and select **send for approval**.

    h. Click **Close**.

    i. On the Inflation Valuation Model page, click **Workflow actions**.

    j. Select **Non-model created**.

    k. Enter a reason for the change and click **Save**.

    l. Sign out of Model Risk Management.

    m. Sign in as **Kfeldman@saspw** using the password **Orion123**.

    n. On her home page, under Tasks, select for **Approve non-model for Model for Inflation Valuation (non-model)**.

    o. Click **Workflow actions**.

    p.   Select **Approve**.

    q.   Sign out of Model Risk Management.

    End of Demonstration

Model Inventory

A key requirement of SR 11-7 is to maintain an accurate and complete model inventory with a comprehensive set of information about models across all of its life stages. The model inventory should contain a consolidated record of all models with related risks, inputs, and outputs, intended uses, controls and stakeholders. In addition to tracking a model individually, a well-maintained model inventory can help in evaluating a firm's aggregate model risk.[2]

Organizations that create and use models require a systematic way to enter and store information about each model over the course of its entire life cycle. This collection of model entries, including its relationships with other elements of the model risk management framework and its related documents, is called a *model inventory*.

Figure 3.3 shows the model inventory process.

Figure 3.3: Model Inventory Process

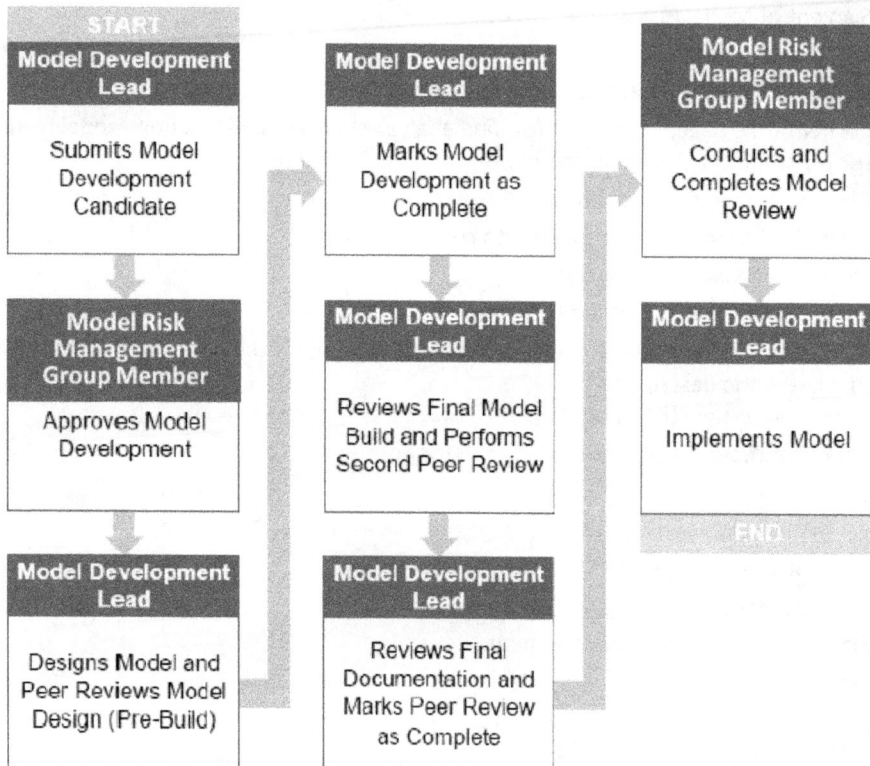

Opening a model from the inventory list reveals the following:

- a model ID as well as the name and description of the model
- the names of model stakeholders, including model developers, model owners, model users, and MRMG members
- links to other model risk management objects, including links to related models, model reviews, model usage information, findings, and action plans
- source code and supporting documentation of the model, which can include model limitations, assumptions, data sources, and so on
- comments about the model
- workflow activity information
- model performance tracking reports

Demo 3.3: Viewing the Model Inventory

1. Submit the model development candidate.
   a. Sign in as **chunter@saspw** using the password **Orion123**.
   b. Click the **Model Inventory** button from the navigation bar.
   c. On the **Model Subcategory** tab, click **New** to create a new model.
      1) For **Classification**, click **Manage**.
      2) Expand **Business Unit** to **Sunshine Financial** ▶ **Wholesale Banking**.
      3) Expand **Model Families** to **Scorecard** ▶ **Credit** ▶ **Retail** ▶ **EAD**.
      4) In the **Model name** field, enter **Retail Credit EAD Scorecard**.
      5) Leave the description blank.
      6) For **Materiality level**, select **Moderate**.
      7) For **Criticality level**, select **High**.
   d. Ensure that the following exist in the Model Stakeholders:
      1) Model Developer – **Carrie Hunter**
      2) Model Owner – **Carrie Hunter**
      3) CoE Coordinator – **Paul Melhoff**
      4) Model User – **Ann Kellog**
      5) MRMG Group Member – **Kirsten Feldman**
      6) OpRisk Manager – **Raul Nunzio**
   e. Click the Model Metadata tab.
      1) For **Is this a Non-Model?,** select **No**.
      2) For **Vended Type**, select **Bank Developed**.
      3) Select **No** for model to be used before validation.
   f. Click the **Documentation** tab.
      Verify that the documents are attached. These are just placeholders.
   g. Click **Save** to save the model.

    h.   From the Workflow actions, select **Send to MRMG** to send the model for approval.

    i.   Sign out.

2.   The MRMG approves model development.

    a.   Sign on as **Kfeldman@saspw** using the password **Orion123**.

    b.   From the Tasks table on Kirsten's home page, select **Retail Credit EAD Scorecard**.

        1) Click **Model Metadata**.

        2) Enter **03/01/2019** for **Date Entered in Inventory**.

    c.   From the Workflow actions, select **Approve Model Development**.

    d.   Sign out.

3.   The model developer designs model design (pre-build).

    a.   Sign on as **chunter@saspw** using the password **Orion123**.

    b.   From the Tasks and Notifications table, select **Retail Credit EAD Scorecard**.

    c.   Click the **Development Details** tab.

        1) For **percent complete**, enter **26%-50%**.

        2) Select **Yes** for **Is conceptual soundness documented**.

        3) Select **No** for **Is peer review complete**.

    d.   From the Workflow actions, select **Peer Review #1 Completed**.

    e.   Enter a change reason.

    f.   Click **Save**.

    g.   Sign out.

4.   The model developer completes model development.

    a.   Sign on as **chunter@saspw** using the password **Orion123**.

    b.   From the Tasks table, select the model **Retail Credit EAD Scorecard**. Make note that the activity is Develop Model.

    c.   Click the development **Details** tab.

        Under the Development Progress section, select **Completed** for **Percent Complete**.

    d.   Click **Workflow actions**.

        1) Select **Development Complete**.

        2) Enter reason text and click **Save**.

        3) Sign out.

5.   The model developer marks second peer review complete.

    a.   Sign on as **chunter@saspw** using the password **Orion123**.

    b.   From the Tasks table, click **Retail Credit EAD Scorecard**.

    c.   Click the **Development Details** tab.

        Under the Development Progress section, enter **I have completed a second peer review of the model under development and reviewed finalized source code and test has verified results.** for the Peer Review (post-build) comments.

      d.   Click **Workflow actions**.

          1) Select **Peer Review #2 Completed**.

          2) Enter a change reason.

          3) Click **Save**.

      e.   Sign out.

6.   Mark peer review as complete.

      a.   Sign in as **chunter@saspw** using the password **Orion123**.

      b.   From the Tasks table, click the model **Retail Credit EAD Scorecard**.

      c.   Click the **Documentation** tab.

          1) For **Is model development documentation attached?**, select **Yes**.

          2) In the Documentation Checklist, select the following:

              a) **Complete Model Documentation Template**

              b) **Methodology Description**

              c) **Testing Documentation**

              d) **Relevant Policies**

              e) **User Manuals**

      d.   Click the development **Details** tab.

          1) For **Is Peer Review Complete?**, select **Yes**.

          2) Enter **I have completed the review of the model documentation.** for the Peer Review (post-doc) comments.

      e.   Select **Workflow actions**.

          1) Select **Peer Reviews Completed**.

          2) Enter a reason text.

          3) Click **Save**.

      f.   Sign out.

The next step is model review, which is completed in the next chapter.

End of Demonstration

Model Data Sources

Data source management is an important element in managing model risk. Models rely on accurate, up-to-date sources of information. Data in banks is often captured in multiple systems, and inadequate system controls can lead to data integrity issues. Models that rely on this faulty data are fundamentally flawed. Therefore, the process of capturing and tracking model data sources is integral to model risk management.

During model development, model developers make decisions about which data sources the model is built upon. Different types of data sources include the following:

- input (transactional) data
- scenario data
- estimates, forecasts, and predicted data

Each data source can have unique properties. For example, some data sources are static (that is, the data source does not regularly change) and other data sources are dynamic (the data changes regularly). Some model-related data is refreshed regularly and other model-related data is refreshed annually or on an ad hoc basis. Some data is internal (bank developed) and other data, such as consortium data, is vended.

Each data source is typically maintained by the following roles:

- Data Owner
- Data Steward
- Subject Matter Expert
- Quality Analyst

SAS Model Risk Management has a data source management feature that enables users to inventory their model data sources and enter metadata about them. These parties and other model governance personnel can then review, track, and report on these data sources.

Model Variables

Model variables refer to the inputs that are used by models to calculate results. Because models are often complex algorithms, they often use a number of variables derived from other sources.

For example, your model might calculate expected loss (EL) for each borrower in a credit risk model as the function EL=PD×LGD×EAD. In this function,

- PD is the probability of default,
- LGD is the loss given default, and
- EAD is the exposure at default.

In this example, the probability of default on a loan might be derived from a company's credit rating. A company with a AAA rating might have only a 0.6% probability of default, whereas another company with a B rating might have a 40% probability of default. In this case, the credit rating might be considered another variable from which PD is derived. Any number of variables can be entered into the SAS Model Risk Management system.

You can draw relationships from these variables to a model by linking variables to one or more models in your model inventory. Other variables might be properties of certain assets or classes of assets in an organization's portfolio. For example, typically exposure at default (EAD) is defined as the outstanding debt at time of default, which is often the balance of the debt.

---

[1] F. Reserve, "Supervisory guidance on model risk management", Board of Governors of the Federal Reserve System, Office of the Comptroller of the Currency, SR Letter, pp. 11-7, 2011.
[2] Ibid

Chapter 4: Model Review, Findings, and Action

Model Review

Models must be reviewed to ensure they perform adequately and meet ongoing operational and regulatory requirements. Model reviews pertain not only to new models that have been developed and need validation but also to existing models that must be reviewed, monitored, and re-validated periodically. The SAS Model Risk Management model review workflow is designed to facilitate these review requirements.

The model review life cycle is controlled by a workflow specification that can be easily modified to conform to an organization's business processes. There are a number of model review types, including the following activities:

- annual review
- full scope validation
- model monitoring and maintenance
- model performance review

> **Note**: The Model Review Validation Rating must be entered into the most recent Model Risk Assessment to automatically flow the model object.

Figure 4.1 shows the model review process is Model Risk Management.

Figure 4.1: Model Review Process

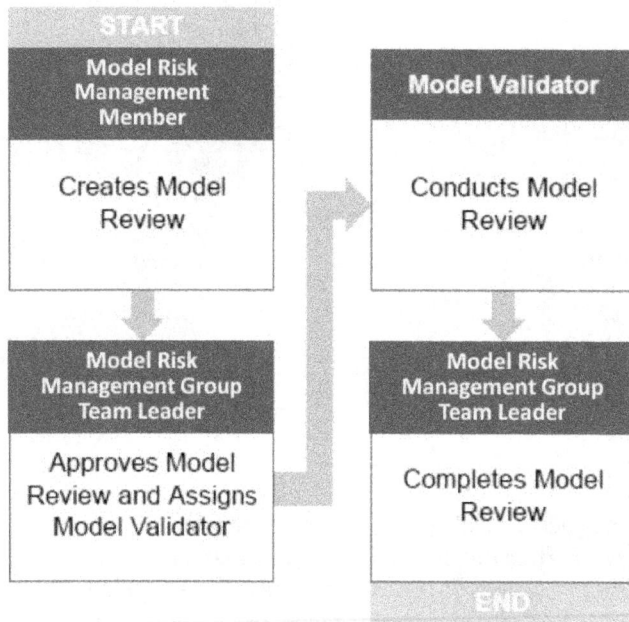

Demo 4.1: Processing a Model Review

1. Create the model review.
   a. Sign in as **Kfeldman@saspw** using the password **Orion123**.
   b. Click the model **Review Category** button.
   c. Click **New** to create a new Model Review.
   d. For **Classifications**, select **Manage**.
      1) For **Business unit**, select **Sunshine Financial ▶ Wholesale Banking**.
      2) For **Review Type**, select **Full Scope Validation**.
      3) For **Model Family**, select **Scorecard ▶ Credit ▶ Retail ▶ EAD**.
   e. On the Model tab, click **Add Link** to link the Retail Credit EAD model.
      1) Select the box for the **Retail Credit EAD model**.
      2) Click **OK**.
   f. In the model review **Details** section, enter the following:
      1) In the **Title** field, enter **Full scope validation of the Retail Credit EAD**.
      2) Select **03/15/2019** as the planned start date.
      3) Select **04/15/2019** as the planned end date.
   g. In the MRMG Group Member table, ensure that **Jeff Estroff** is listed. You can also remove the others if you choose.
   h. Click **Save** to save changes to the model review.

      i.   From **Workflow actions**, select **Submit**.

      j.   Sign out.

2.   The MRMG team leader assigns a model validator.

    a.   Sign in as **jestroff@saspw** using the password **Orion123**.

    b.   From the task list, select **Triage full scope validation process for Full scope validation of Retail Credit EAD**.

    c.   On the Details tab, go to the Model Validator table.

        1) Click **Add Link**.

        2) Select **Greg Telfer**.

        3) Click **OK**.

    d.   Save the model review.

    e.   From **Workflow actions**, select **Submit for Validation**.

    f.   Sign out.

3.   The model validator conducts a model review and provides validation documentation.

    a.   Sign in as **gtelfer@saspw** using the password **Orion123**.

    b.   b. From the Task table on his home page, **click Full Scope Validation of the Retail Credit EAD**.

    c.   Go to the review-specific Details tab.

        1) The model has high model risk materiality. Select **1-High** for **MRMG Materiality**.

        2) The model is ready for use, so click **Fit for Intended Use with Conditions in the Validation Outcome field**.

        3) Enter **09/15/2019** as the next validation date.

        4) For **Issue management done**, select **No**.

    d.   Click **Workflow actions** and select **Submit**.

        1) Enter a change reason.

        2) Click **Save**.

    e.   Sign out.

4.   Complete the model review.

    a.   Sign in as **jestroff@saspw** using the password **Orion123**.

    b.   From the Tasks table, click **Full scope validation of the Retail Credit EAD**. Notice that the action is **Go to model**, and mark **Validation Complete**.

    c.   The model is reviewed. Click **Workflow actions**.

    d.   Select **Submit**.

    e.   Sign out.

        The model has a finding. From here, the model moves to Findings and Action Plans to deal with the conditions brought up by the model validator. This is covered later in this chapter.

End of Demonstration

Model Monitoring

Model Monitoring is a content package for SAS Infrastructure for Risk Management that integrates the inventory and management capabilities of SAS Model Risk Management. The Model Monitoring tool in SAS Model Risk Management centralizes model-performance calculations, performance tracking and alerts, model data storage, and SAS Visual Analytics reporting.

In SAS Infrastructure for Risk Management, an administrator creates a new instance in the model monitoring flow category. While the model monitoring flow executes, the model is scored and data is loaded to the Reporting Mart and the LASR Analytic Server. The model monitoring flow evaluates models based on the following measures:

- accuracy ratio (Gini)
- chi-square test (*p*-value)
- Kolmogorov-Smirnov statistic
- system stability index
- mean squared error

When the flow executes, a model monitoring object is created in SAS Model Risk Management. Model monitoring assesses three thresholds: threshold low, threshold high, and threshold outliers. Any model with a measured value that is less than threshold low, greater than threshold high, or equal to threshold outliers generates a finding for the object as shown in Figure 4.2. These thresholds are defined by an administrator.

Figure 4.2: Model Risk Management Technical Design for Model Monitoring

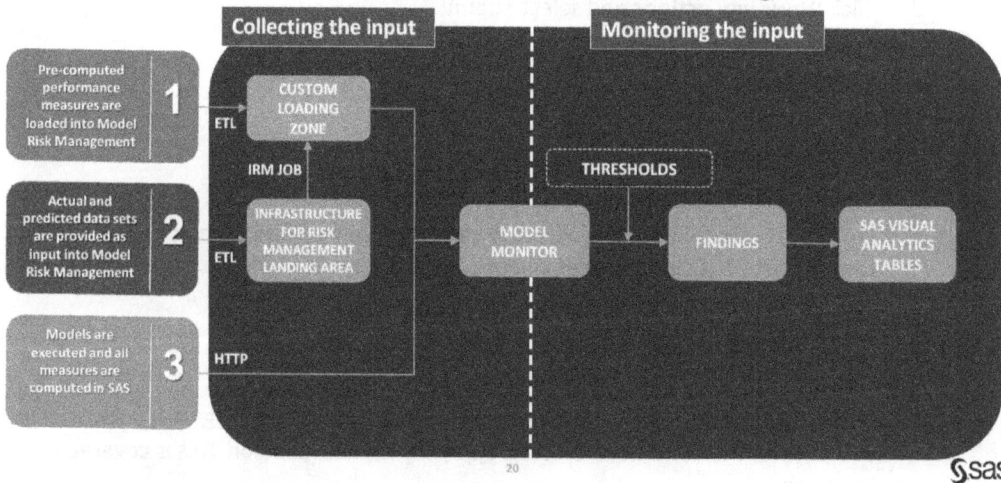

Findings and Actions

In model risk management, *findings* are issues noticed in the course of a model review or validation that present weaknesses or shortcomings in the model. Findings could include problems identified in the model documentation, computational issues, or data limitations.

Action plans are concrete measures typically developed to mitigate or respond to findings, although they might be stand-alone tasks created in response to other situations.

The life cycle management of findings and action plans is achieved through discrete stages as shown in Figure 4.3. The process requires accountability and creates a traceable information stream for auditing and reporting purposes.

Figure 4.3: Findings and Action Plan Process

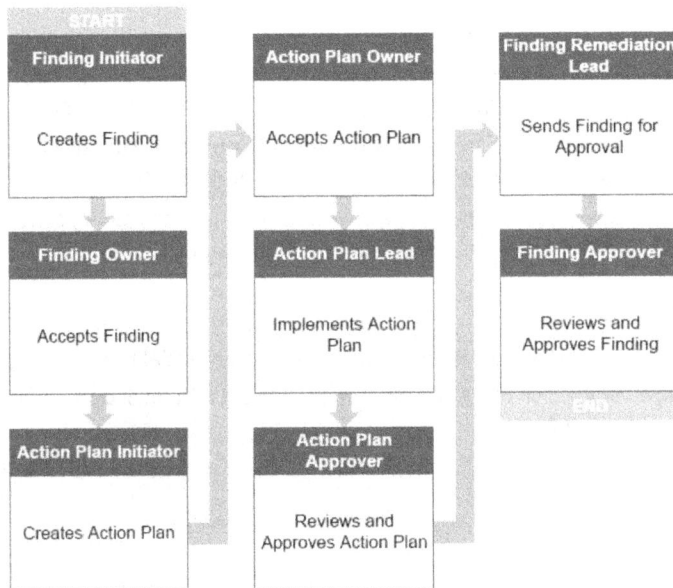

Findings and action plans have separate stages and workflows, but these stages are integrated. The separation of each workflow enables flexibility around the discovery and mitigation process. For example, you can create multiple action plans for a single finding or respond to multiple findings through a single action plan.

Demo 4.2: The Finding and Action Plan Process

1. Create a finding.
   a. Sign in as **jestroff@saspw** using the password **Orion123**.
   b. Click the **Findings** category button.
   c. In the Findings table, select **New** to create a new finding.

    d.   For **Classification**, click **Manage**.

        1) For **Business Units**, select **Sunshine Financial ▶ Wholesale Banking**.

        2) For **Model Families**, select **Scorecard ▶Credit ▶ Retail ▶ EAD**.

        3) For **Review Type**, select **Full Scope validation**.

        4) Click **Save Point**.

    e.   On the Details tab:

        1) Enter **Potential model risks with Scorecard Credit Retail EAD model** as the title.

        2) Enter **The model risks with the scorecard model have not been identified or assessed** for the description.

        3) Select **High** for **Severity Tier**.

        4) Select **04/01/2019** for **Current Due Date**.

        5) Select **Process improvement** for **Category**.

        6) Jeff initiated the finding. Ensure that an initiator has been selected.

        7) Add a remediation lead. In the Remediation Lead table, click **Add Link**. The Add Link window appears. Select **Carrie Hunter** as the remediation lead and click **OK**.

        8) Add a finding owner. In the Owner table, click the **Add Link** button. The Add Link window appears. Select **Carrie Hunter** as the finding owner and click **OK**.

        9) Add an approver. Kirsten will serve as the approver. In the Approver table, click **Add Link**. The Add Link window appears. Select **Kirsten Feldman** as the finding approver and click **OK**.

    f.   Click the **Linked Objects** tab.

        1) Click **Add Link**, and select the **Credit Retail EAD** model in the Model table.

        2) In the Model Review table, click the **Add Link** button and select **Full Scope Validation** for the credit retail EAD.

    g.   Click **Save**.

    h.   Click **Workflow actions**.

    i.   Select **Publish**.

    j.   Sign out.

2.   The finding owner accepts the finding.

    a.   Sign in as **chunter@saspw** using the password **Orion123**.

    b.   From the task list, click the **Accept Finding for the Credit Retail EAD** model.

    c.   On the Details tab, enter **Conduct a review to create any model risks associated with this model, and conduct a risk assessment.** for Recommended/Resolution.

    d.   Click **Workflow actions**.

    e.   Select **Accept**.

    f.   Sign out.

3.   The action plan initiator creates an action plan for the finding.

    a.   Sign in as **chunter@saspw** using the password **Orion123**.

    b.   From the Tasks table, click the process finding for the **Credit Retail EAD** model.

    c.   On the Linked Objects page in the Action plan table, select **New** to create a new action plan.

d. On the Action Plan page, on the Details tab, do the following:

1) Enter **Identify risks and conduct risk assessment** for the title.

2) Enter **Review the model to identify model risks and conduct a risk assessment on the Scorecard model.** as the action plan description.

3) Select **High** for **Priority**.

4) Select **4/30/2019** for **Due Date**.

5) If Carrie is not currently selected as the initiator, add an action plan initiator by clicking **Add Link,** selecting Carrie, and clicking **OK**.

6) Add a lead. In the Lead table, click the **Add Link** button, select **Carrie Hunter**, and click **OK**.

7) In the Owner table, click the **Add Link** button, select **Carrie Hunter**, and click **OK**.

8) If Carrie is not already the action plan owner, click the **Add Link** button, select **Carrie Hunter**, and click **OK**.

9) For **Approver**, click the **Add Link** button. Select **Kirsten Feldman** and click **OK**.

e. Go to the **Linked Objects** tab.

1) In the Model table, click **Add Link** and select the **Scorecard Credit Retail EAD** model.

2) In the Model Review table, select **Full scope validation Credit Retail EAD**.

f. Click **Save**.

1) Add a change reason if necessary.

2) Click **Save**.

g. Click **Workflow actions**.

h. Select **Publish**.

i. Sign out.

4. The action plan owner accepts the action plan.

a. Sign in as **chunter@saspw** using the password **Orion123**.

b. From the task list on her home page, click **Accept Action Plan** for the **Identify risks and conduct risk assessment** action plan.

c. Click **Workflow actions**.

d. Select **Accept**.

e. Sign out.

5. The action plan lead implements the action plan.

a. Sign in as **chunter@saspw** using the password **Orion123**.

b. From the task list on the Home page, click the **Implement Action Plan** activity for the **Identify risks and conduct risk assessment** action plan.

c. On the Details tab, select the current date for **Action Plan Closed Date**.

d. Click **Workflow actions**.

e. Select **Send for Approval**.

f. Sign out.

6. The action plan approver reviews and approves the action plan.
   a. Sign in as **kfeldman@saspw** using the password **Orion123**.
   b. From the task list on the Home page, click the **Approve Action Plan** activity for the **Identify risks and conduct risk assessment action plan.** The Edit Action Plan page appears.
   c. On the Details tab, enter **The model risk assessment was completed as requested.** for the Independent Model Validator assessment for action plan.
   d. Click **Workflow actions.**
   e. Select **Approve**.
   f. Sign out.
7. The finding remediation lead sends the finding for approval.
   a. Sign in as **chunter@saspw** using the password **Orion123**.
   b. From the task list on the Home page, click the activity **Process Finding for Potential model risks with Scorecard Retail EAD.**
   c. Click **Workflow actions.**
   d. Select **Send for Approval**.
   e. Sign out.
8. The finding approver reviews and approves the finding.
   a. Sign in as **kfeldman@saspw** using the password **Orion123**.
   b. From the task list, click the activity approve finding for **Potential model risks with Scorecard Retail EAD.**
   c. Click **Workflow actions.**
   d. Select **Approve**.
   e. Sign out.
9. The model validation cycle is complete.
   a. Sign in as **kfeldman@saspw** using the password **Orion123**.
   b. In the task list, click the activity In**itiate and conduct validation for Retail Credit EAD Scorecard.**
   c. Click **Workflow actions.**
      1) Select **Model Validation cycle complete.**
      2) Enter reason text and click **Save**.
   d. Sign out.

10. Implement the model.

    a. Sign in as **chunter@saspw** using the password **Orion123**.

    b. From the task list, click **Implement model for the Retail Credit EAD Scorecard**.

    c. Click **Additional metadata**.

        1) Enter today's date for the implementation date.

        2) **For Implemented?**, click **Yes**.

    d. Click **Workflow actions**.

    e. Select **Model Implementation Complete**.

    End of Demonstration

Chapter 5: Model Risk Assessment and Attestation

Model Risk Assessment

Model risk assessment is an activity focused on specifying and quantifying model risk. Model risk quantification uses model risk factors such as the following:

- model usage (why is the model being used?)
- model materiality (how much capital is the model directing?)
- model criticality (how important is the model to the business?)

These factors are used to calculate the *inherent risk,* the amount of risk that would exist if no controls were put in place. Inherent risk is used in a secondary calculation, along with the control score of a model, to calculate the overall *residual risk,* the amount of risk after mitigation through controls.

Model Risk Assessment Process

First, model risk tiering is performed. Questions are answered in the following areas:

- model usage
- model criticality
- model materiality

These questions are used to determine the model's risk tier. Tier 1 risks are considered the most material and critical. Tier 4 risks are considered non-material. The following flow chart is used in the solution by default to determine risk tiers.

Figure 5.1: Risk Tier Determination

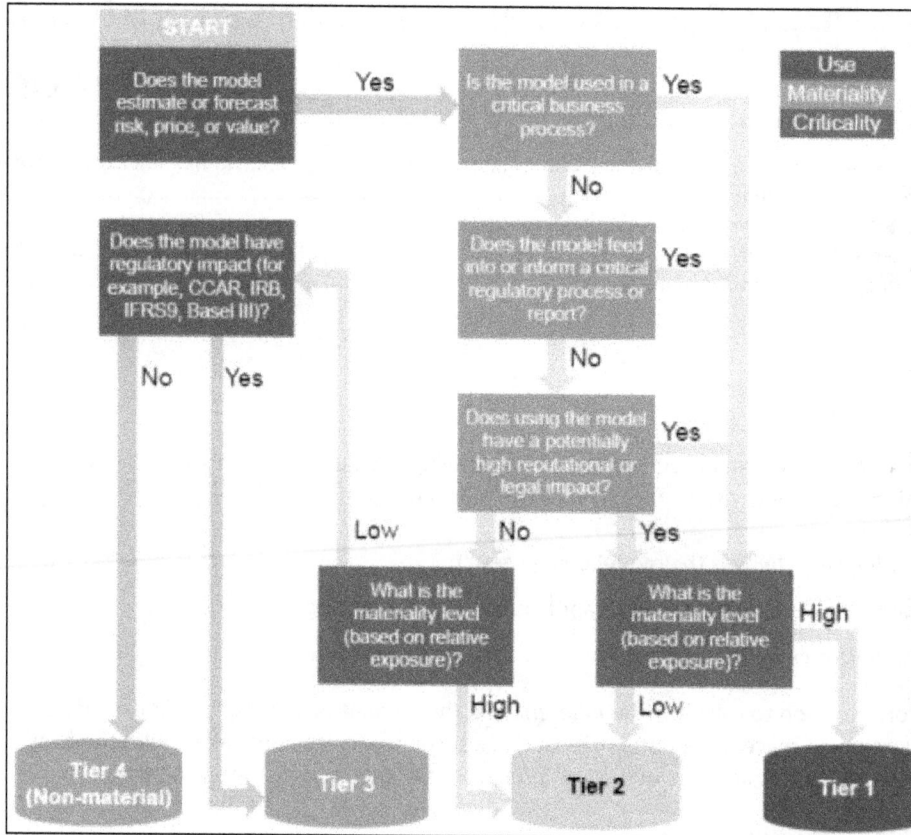

After the risk tier has been established, additional information is taken from the most recent model validation rating. These factors are used to determine the overall inherent risk rating for the model. (See Figure 5.2.)

Figure 5.2: Inherent Risk Score

		Control Environment		
		Strong	Moderate	Weak
Inherent Risk	High	Low	High	High
	Medium	Low	Medium	High
	Low	Low	Low	Medium

Next, control assessment is performed. A second series of questions is answered in the following areas:

- model design
- model use
- regulatory and audit scope

For example, a model design question might be "Is the model methodology in line with the industry standard?" Likewise, a regulatory question might be "Could the model have an indirect impact on regulatory capital?"

These questions determine the control score. Control scores can be low, medium, or high. A model with low controls has a higher level of risk than a model with high controls. The inherent risk score, combined with the control score, is used to formulate the residual risk score.

Figure 5.3: Residual Risk Score

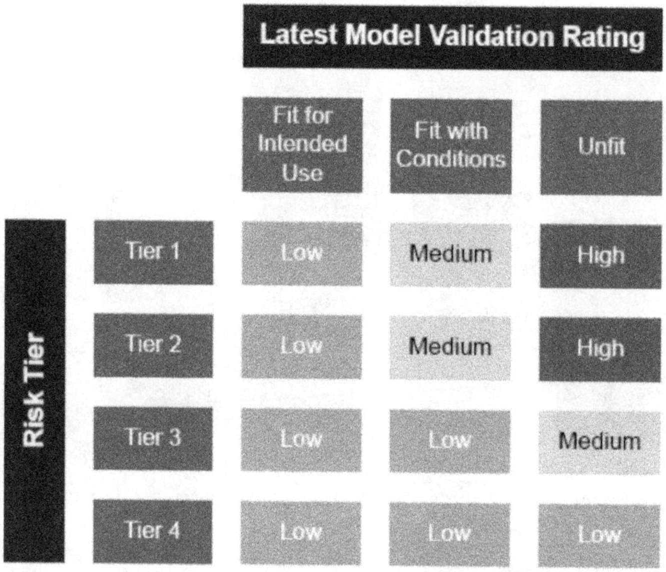

Latest Model Validation Rating		
Fit for Intended Use	Fit with Conditions	Unfit

Risk Tier	Fit for Intended Use	Fit with Conditions	Unfit
Tier 1	Low	Medium	High
Tier 2	Low	Medium	High
Tier 3	Low	Low	Medium
Tier 4	Low	Low	Low

Demo 5.1: Perform a Model Risk Assessment

1. An MRMG member creates a model risk assessment and assesses the model.
   a. Sign in as **kfeldman@saspw** using the password **Orion123**.
   b. On the Model Risk Assessment subcategory tab in the Model risk table, click **New** to create a new model risk assessment.
   c. Expand **Classification**.
   d. Expand Business unit to **Sunshine Financial ▶ Wholesale Banking**. Select **Wholesale Banking**.
   e. Expand Model Families to **Scorecard ▶ Credit ▶ Retail ▶ EAD**.
   f. Click **OK**.
   g. Enter **Model Risk Assessment for EAD Credit Scorecard model** for **Model Risk Name**.
   h. Enter **Assesses the model risk for the EAD Credit Scorecard model** for the description.
   i. Enter **3/20/2019** for **Risk Assessment Start Date**.
   j. Specify all three options for **Model Risk Assessment Type**.
   k. In the Models table, click the **Add Link** button.
   l. Select **Retail Credit EAD Scorecard**.
   m. Click **OK**.

n. Click the **Risk Tier** tab.

1) The selections that you make on the Risk Tier tab are dependent on how you answer questions. Some questions are skipped if they are not relevant to the risk tier calculation. To view a flow chart that explains how your selections impact the risk tier, click the **Risk Tier Methodology** explainer link.

2) Kirsten has determined that the risk meets the following criteria:

a) The model is used to estimate a risk value (Question 1).

b) The model is not used for a critical business process (regulatory reporting) (Question 2).

c) The model is of high materiality (Question 5b).

3) She selects these answers to the questions presented and views the risk tier. Based on these answers, the model should be rated as a Tier 1 model.

o. Return to the **Details** tab.

p. Under the Latest Validation Rating, select **Fit for intended use**.

q. Click the **Control Assessment** tab.

r. Follow the form below for the answers to each question.

A. Assumptions:

A1: Are the modeling assumptions identified by model owners? **Yes**

A2: Does the model heavily rely on certain key assumptions, or is the model output significantly sensitive to changes in assumptions? **Yes**

B. Methodology:

B1: Are the model owners aware of the industry standard for this methodology? **Yes**

B2: Is the model methodology in line with the industry standard? **Yes**

B3: How complex is the model? **Moderately complex**

B4: If applicable, is the financial contract or asset that the model is applied to exotic or complex? **Yes**

C. Documentation:

C1: Is model documentation available? **Yes**

C2: Is the model documentation up-to-date and subject to versioning? **Yes**

C3: If required, is the model documentation up to the regulatory and internal standard? **Yes**

D. Input Data:

D1: Is there a high level of confidence in data inputs with respect to sources, quality, and accuracy? **Yes**

D2: Are there well-defined procedures for manual data preparation and cleaning? **Yes**

D3: Is the data entry automated to limit the human input and errors? **Yes**

D4: Is any of the model input data hard to obtain or unobservable? **No**

D5: Does the model input data need to be sourced externally? **No**

D6: Are there known issues with the data quality (internal and external)? **No**

D7: Does model input data require complex processing? **No**

D8: Are any model inputs based on expert judgment? **Yes**

E. Calibration:

E1: Does the model need parameter calibration during its development? **Yes**

E2: Does the model need frequent parameter calibrations after implementation (more than once a year)? **No**

F. Implementation:

F1: Is the model implemented as a registered EUC or on a strategic system? **Yes**

F2: Is the model implementation reviewed by any independent group? **Yes**

F3: Does the model require complex coding? **No**

F4: Has the model any history of implementation issues? **No**

G. Performance:

G1: Does the model have any history of performance issues? **No**

G2: Does the model have any suspected performance issues? **No**

H. Governance:

H1: Is the model owner team also the model developers / purchasers? **Yes**

H2: Is the model under the review of a governance committee or subject to a specific policy? **Don't Know**

H3: Is there a documented history of model versions? **Yes**

I. Model Monitoring:

I1: Does the model undergo regulator monitoring or re-assessment? **Yes**

I2: Are the model assumptions and limitations reviewed regularly? **Yes**

J. Model Use:

J1: Is a user manual (including formal procedures, guidelines, process documents) available for the model? **Yes**

J2: Are the model users also the model owners/developers? **Yes**

J3: How many users does the model have? **1-10**

K. Financial Impact:

K1: Could the model issues lead to a direct and material financial loss? **No**

K2: Could the model issues lead to an indirect or nonmaterial financial loss? **Yes**

K3: Could the model issues lead to a misstatement of the value of contracts or products? **No**

K4: Could model issues lead to a misstatement of liquidity requirements? **No**

L. Business Impact:

L1: Could model issues lead to a direct impact on funding, hedging, or capital strategies? **No**

L2: Could model issues lead to loss of market share in the business where it is used? **Yes**

    M. Downstream Dependencies:

        M1: Does one or more of the firm's models or tools rely on the accuracy of this model's output? **Yes**

        M2: Are any of the downstream model's material in the sense as outlined above? **No**

    N: Regulatory/Audit Scope

        N1: Could model issues have a direct and material impact on regulatory capital? **No**

        N2: Could the model have an indirect impact on regulatory capital? **No**

    O. Reputational Impact:

        O1: Could model issues lead to an adverse reputational outcome for the firm? **No**

        O2: Is the model output externally published to the market? **No**

    P. Audit Issue:

        P1: Has the model been subject to audit or regulatory review? **Yes**

        P2: Does the model have any history of regulatory or audit issues? **No**

s.   Click the **Details** tab.

t.   Ensure that **Override Scores** is set to **No**.

u.   The values for **Inherent Risk Rating** and **Control Score** determine the value for **Residual Risk Rating**. To view the heat map that is used to determine the residual risk rating, click the **Residual Risk Rating Methodology** explainer link.

v.   Click the **Save** icon. Click **Close** to complete the model risk assessment.

    End of Demonstration

TRIM (Targeted Review of Internal Models)

Banks often rely on internal models to estimate and report on risks such as credit risk, market risk, and operational risk. However, regulatory agencies have raised concerns about an overall lack of trust in those internal models, based on inconsistencies in the calculations used across different banks.

A targeted review of internal models, or *TRIM project*, is designed to help banks prepare for regulatory inspection, particularly in areas such as modeling methodologies, the risk governance framework, model risk management, data quality, and documentation of model life cycle processes. TRIM assessments are intended to reduce unwarranted variables in risk-weighted assets (RWA) and to thereby ensure the consistency, reliability, and comparability of the internal models that banks use to determine regulatory capital compliance.

TRIM projects in SAS Model Risk Management include a series of questionnaires divided into the following topics:

- **Topic 0: Overarching principles.** Topic 0 covers all general organizational principles and guidelines concerning model documentation, life cycle, and validation.

- **Topic 1: Roll-out and PPU.** Topic 1 covers the organization's roll-out plan and permanent partial use (PPU) procedures.

- **Topic 2: Internal governance, internal reporting, and organization of GRCU.** Topic 2 covers the structure, responsibilities, and reporting hierarchies of the organization's management bodies and committees.

- **Topic 3: Internal audit.** Topic 3 covers the organization, frequency, and findings of the organization's internal auditing framework.

- **Topic 4: Internal validation.** Topic 4 covers the functions, duration and frequency, policies, depth, and breadth of the organization's internal validation framework.

- **Topic 5: Modes use.** Topic 5 covers the management, rating, approval processes, allocations, functions, and overrides of an organization's internal models.

- **Topic 6: Management of model changes and extensions.** Topic 6 covers the documentation and processes of an organization's management of model change.

- **Topic 7: Data-quality management.** Topic 7 covers the processes, collection requirements, and infrastructure of an organization's data-quality management.

- **Topic 8: Third-party involvement.** Topic 8 covers the tasks that an organization outsources to a third party as well as the types of models outsourced and internal, third-party performance control and monitoring processes.

- **Topic 9: Assignment of exposures to exposure classes.** Topic 9 covers the processes (general, manual, and automatic) concerning an organization's assignment of exposures to exposure classes.

- **Topic 10: Default definitions.** Top 10 covers the organization's definitions and indications of default as well as the organization's processes to anticipate, predict, recognize, and mitigate default.

Each topic includes a Supporting Documentation table to link relevant documentation to the TRIM assessment for internal auditing purposes. A topic's Supporting Documentation table does not accept multiple uploads of documents with the same file name. However, users might want to upload the same document multiple times to support multiple questionnaire responses.

Uploading a single document multiple times might be especially useful for long documents in which a particular section supports a response to one question and another section of the same document supports a response to a different question. A document can be uploaded multiple times so long as each upload uses a different file name.

One approach would be to name supporting documents based on the topic, question, and sub-question that it answers, as well as the section of the document that supports that answer.

For example, section 12B of a document titled "EXAMPLE" might support a response to sub-question 2 of question 4 under topic 1. In addition, section 4H of that same document might support a response to subquestion 3 of question 2 of topic 1. This document can be uploaded as both T1Q4SQ2_EXAMPLE_S12B" and as T1Q2SQ3_EXAMPLE_S4H. In this example, "T" represents the topic, "Q" represents the question, "SQ" represents the sub-question, "EXAMPLE" is the title of the document, and "S" represents the referenced section of the document.

Figure 5.4: TRIM Process

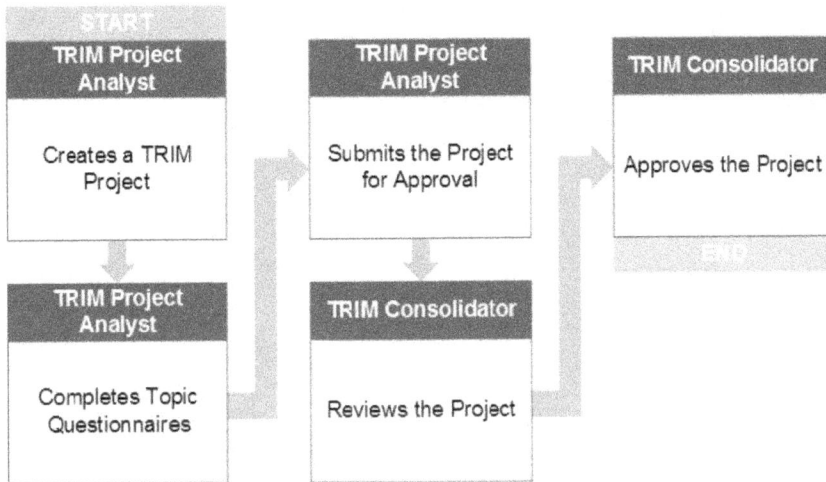

Model Inventory Attestation

Risk managers and senior management need processes and policies that maintain regular oversight over any models under management. This includes the process of inventory attestation, in which model governance members periodically sign off and attest to the accuracy and integrity of the models in their inventory. (See Figure 5.5.) Inventory attestation is typically performed on a set of models that are grouped together for a common purpose (for example, the models are involved in the creation a regulatory report).

Figure 5.5: Inventory Attestation Process

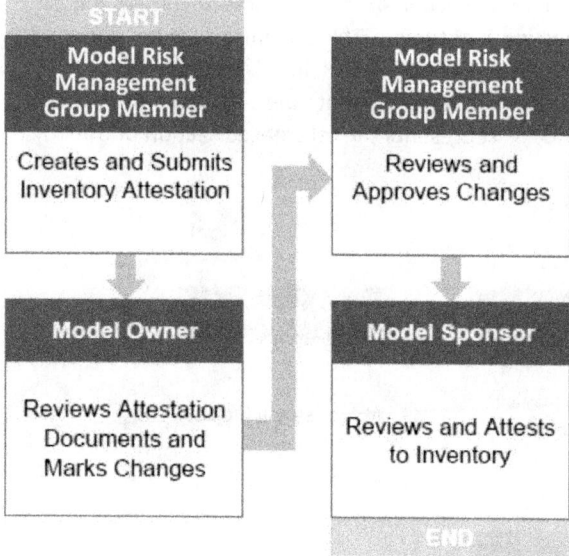

Chapter 6: Model Usage and Change Management

Model Usage

After a model has been added, reviewed, and implemented, model users can begin running and generating output from the model. To use an existing, implemented model, users can submit model usage information to model owners in a model usage form. This might include details about their usage and can include their business justification for running the model, with model owners reviewing those requests and providing approvals. In addition, users can begin using the model and use a model usage to provide feedback on their experience with the model.

After model owners have approved the usage, the MRMG also reviews the model usage documentation and determines whether additional assessment is needed, providing an additional line of defense. This model usage information can be tracked and used to generate reports for use by the MRMG, regulators, auditors, and other interested parties. After the model usage has been approved, steps can be taken outside SAS Model Risk Management to provide access to the model.

Model owners can also create a model usage to inform model users of the availability of the model, to document the model's expected usage, and to invite feedback. Model users can review the model owners' invitation to use the model, provide feedback, and approve the model usage to accept the model as a valid tool for use. Alternatively, model users can reject the model usage and send it back to the model owner to indicate that the model is insufficient to meet their needs. Similar to the first scenario, the MRMG can review the usage and determine whether additional assessment is required.

Figure 6.1: Model Usage Process

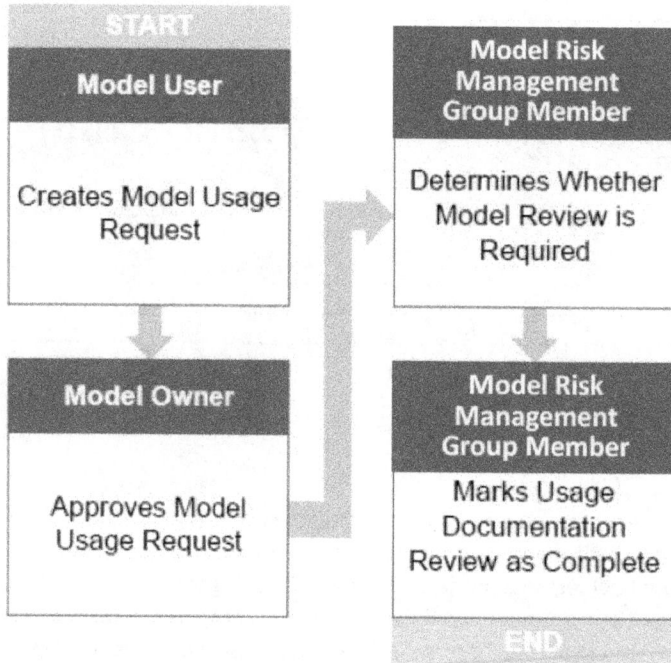

Model Change Management

In model risk management, change management is a process in which inadequacies in the model domain are identified and steps are taken to correct these inadequacies. Changes can be necessary for a number of reasons, including the following:

- changing the model's algorithms (to account for regulatory changes, fix existing algorithms, modify underlying variables or data, or improve performance)
- requesting a change or addition to the usage of the model
- requesting an exception

It is important when planning a change to a model to determine whether the change is material—that is, whether the change to the model is significant. Material changes require more extensive documentation and review. The determination of materiality must be made on a case-by-case basis.

Change management incorporates a number of different tasks in order to make changes to an existing model:

- planning and scoping a change to an existing model
- submitting a change request

- reviewing and approving the change request
- implementing the change
- reviewing and evaluating the results of the change

Change management typically involves changes to an existing model to correct behavior, but it can also include model creation.

Figure 6.2: Change Management Process

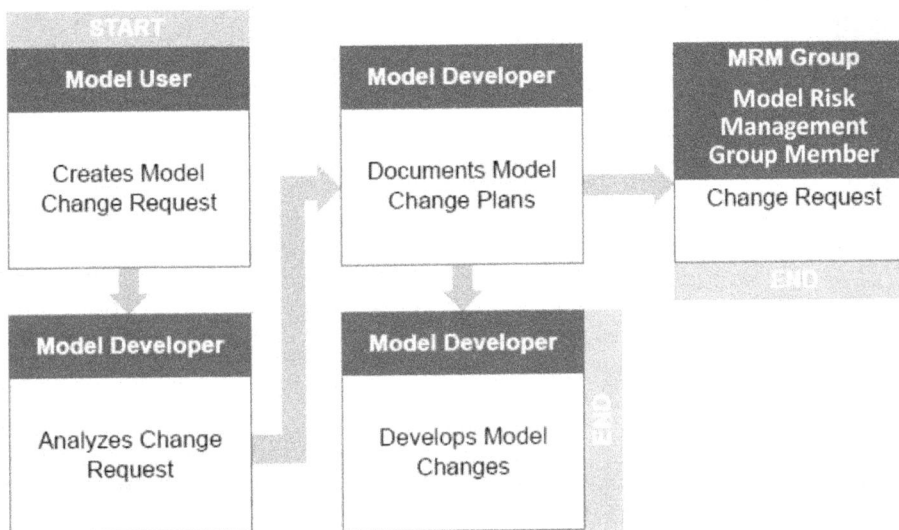

Policy Exceptions

Policies refer to rules or guidelines that govern how an organization executes its operations. In some cases, policy exceptions are required to address contingencies in which a compliance exception might be required. Policy exceptions must be effectively managed to avoid these exceptions becoming the rule. In many instances, policy exceptions should be written into subsequent versions of the policies that they are created to address. The objective of policy exception management in model risk management is to define policy exceptions for the organization and manage their life cycle. The following figure shows how policy management fits into the overall Model Risk Management life cycle.

Figure 6.3: Model Risk Management Life Cycle

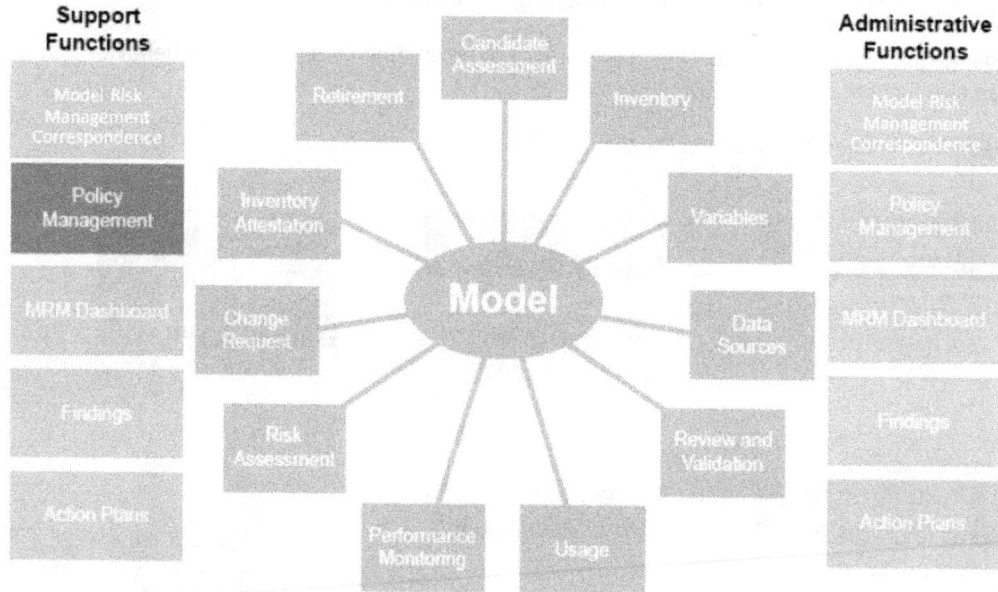

Policy exceptions should be designed to meet certain regulatory obligations and organizational objectives. Policy exceptions can be associated with a variety of business objects, such as models, model candidates, or model risk management controls. For example, an organization might have a policy that defines how model candidates are reviewed and approved before they are used. This policy is intended to ensure that a model candidate is reviewed by the correct MRMG member before going into production. However, due to a backlog in the review process, a critical model cannot be used because it goes against the review policy. A policy exception could be created to temporarily approve the usage of a new model before it has undergone thorough review.

The policy and policy exception life are controlled by a workflow specification that can be easily modified to comply with an organization's business processes. These workflows enable accountability and create a traceable information stream for auditing and reporting purposes.

Figure 6.4: Policy Exception Process

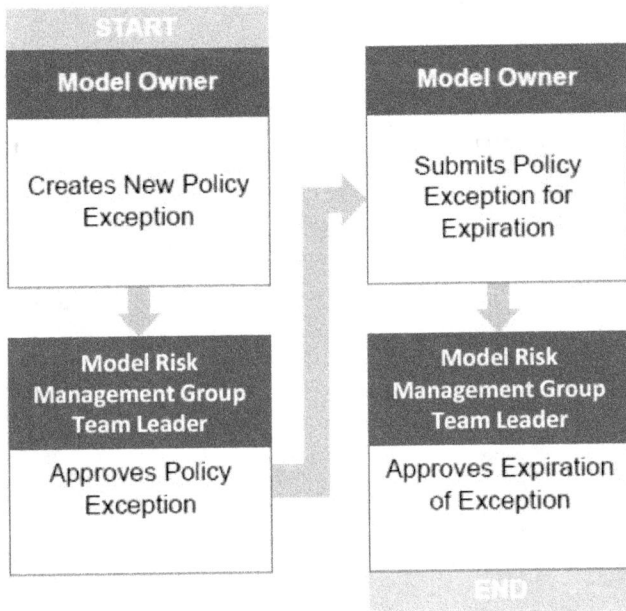

Chapter 7: Additional Topics

Correspondence

Correspondence can be used for a number of reasons:

- to communicate day-to-day activities between groups (for example, between model developers, model users, and the MRMG)

- to enable the modeling team to communicate between processes

- to raise and discuss model development issues to determine whether a finding or action plan is required

- to link various model-related discussions to models, model reviews, model changes, findings and action plans, and so on

- to attach documents, source code, and other files to these communications

- to account for exception situations or rarely occurring situations that might not be captured in the workflow

Having a way to capture these correspondences within the system, rather than using email to communicate, enables the organization to have a central repository that can be reviewed and reported on by auditors, risk managers, and regulators.

Figure 7.1: Correspondence Process

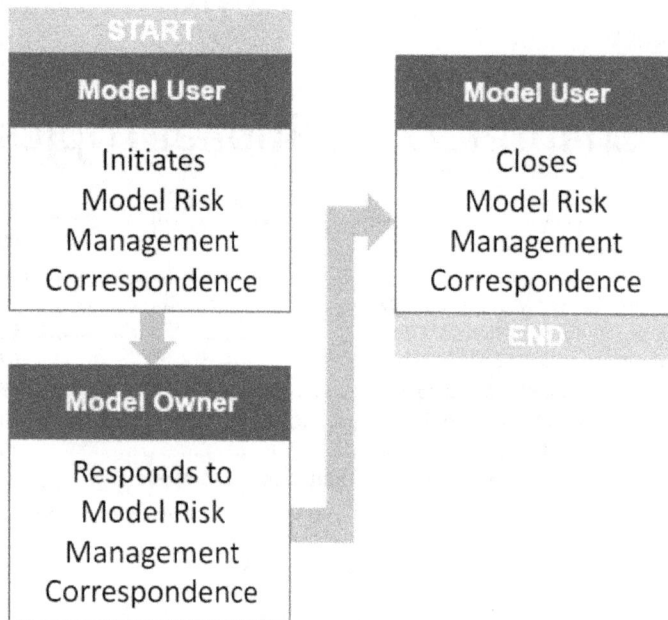

Demo 7.1: Using Correspondence

1. The model user initiates MRM correspondence.
   a. Sign in as **akellog@saspw** using the password **Orion123**.
   b. Click the **Correspondence** category button.
   c. Click **New** to create a new correspondence.
   d. Expand the **Classification** selector.
      1) Expand **Business Units** to **Sunshine Financial ▶ Wholesale Banking**.
      2) Expand **Model Families** to **Scorecard ▶ Credit ▶ Retail ▶ EAD**.
      3) Click **OK**.
   e. Enter **Cannot log on to server to run Retail Credit EAD Scorecard** for **MRM Correspondence Name**.
   f. In the Recipients table, click the **Add Link** button.
      1) Select **Carrie Hunter**.
      2) Click **OK**.
   g. In the CC List Entries table, click **Add Link**.
      1) Select **Jeff Estroff**.
      2) Click **OK**.
   h. Click the **Linked Objects** tab.
   i. In the Model table, click the **Add Link** button.
      1) Select **Retail Credit EAD Scorecard**.
      2) Click **OK**.

      j.   Click **Save**.

      k.   Click **Workflow actions**.

          Select **Send to Recipients**.

      l.   Sign out.

2.   The model owner responds to the MRM correspondence.

      a.   Sign in as **Chunter@saspw** using the password **Orion123**.

      b.   From the Task list, click **Add Comments (Recipient)** for the Retail Credit EAD Scorecard model correspondence.

      c.   Carrie Hunter has also tested this problem and cannot log on to the server that runs the scorecard. She notices that after a recent outage, the server was not turned back on. She reboots the machine and can now log on and run the scorecard.

      d.   On the Details tab in the Messages table, click **Add Comment**.

          1) Enter **The Scorecard server was down after an outage. I have rebooted. Please retry.**

          2) Click **OK**.

      e.   Click **Workflow actions**.

      f.   Select **Notify Initiator and Recipients**.

      g.   Sign out.

3.   The model user closes MRM correspondence.

      a.   Sign in as **akellog@saspw** using the password **Orion123**.

      b.   From the Task list, **click Close Correspondence activity for the Retail Credit EAD Scorecard model**.

      c.   Ann attempts to log on and is successful.

      d.   On the Details tab in the Messages table, click **Create a new message**.

          1) Enter **The scorecard server appears to be working now**.

          2) Click **OK**.

      e.   Click **Workflow actions**.

      f.   Select **Close**.

      g.   Sign out.

End of Demonstration

Dashboards for Management Reports

Governance of the model life cycle is an important and necessary part of the model risk management framework. Senior management and the governing body oversee the entire model risk management operation. They are responsible for determining the level of risk appetite that the organization deems acceptable, reviewing the current status of the operation, planning and strategizing for the future, and making the appropriate decisions in real time. This effort requires a constant stream of up-to-date information.

Figure 7.2: Model Life Cycle Governance

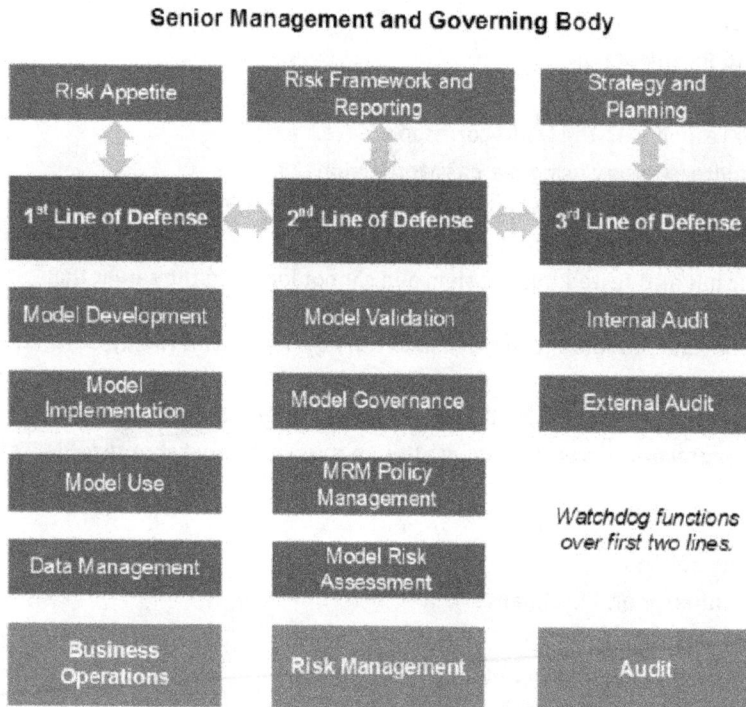

Senior Management and Governing Body

Risk Appetite	Risk Framework and Reporting	Strategy and Planning
1st Line of Defense	2nd Line of Defense	3rd Line of Defense
Model Development	Model Validation	Internal Audit
Model Implementation	Model Governance	External Audit
Model Use	MRM Policy Management	*Watchdog functions over first two lines.*
Data Management	Model Risk Assessment	
Business Operations	Risk Management	Audit

With the right information at the right time, senior managers can do the following:

- quickly assess the overall model risk management picture from top to bottom
- track and schedule objects, such as findings, action plans, model performance, and model reviews
- explore which areas have problems and which areas are functioning well
- discover where delays and bottlenecks exist in the life cycle of models and other model-related objects
- challenge existing practices
- review and approve policies and policy exceptions
- develop and prioritize action items
- allocate resources
- mitigate risks
- communicate with regulators and other internal stakeholders

SAS Model Risk Management achieves information delivery to senior managers and other interested parties through the use of the MRM dashboard. The MRM dashboard is a powerful set of dynamic reports that show the overall state of your model ecosystem. It enables you to interactively explore and drill into your model risk management information, from high-level graphs and figures down to individual models, findings, action plans, and so on.

The MRM dashboard is designed to work with tablets as well as browsers for portability and ease-of-use. The reports within the dashboard can also be imported into Microsoft Office documents, and those documents can be easily refreshed with current information. Reports can also be edited and enhanced by power users.

Data for the reports flows from the SAS Model Risk Management database to the SAS Visual Analytics LASR Analytic Server, which stores the data in memory, as shown in the following figure:

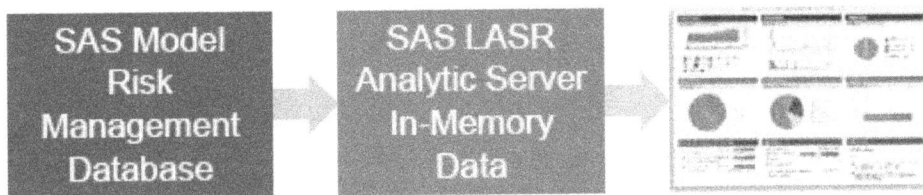

Administrators are responsible for scheduling ETL jobs to regularly update the report information. For more information about scheduling and running jobs to update the report data, see the "Report Administration" chapter in *SAS® Model Risk Management: Administration and Customization Guide*.

Access to the dashboards and data is controlled at an individual level and is dependent on the positions that have been assigned to a particular user.

Figure 7.3: Model Risk Management Dashboard

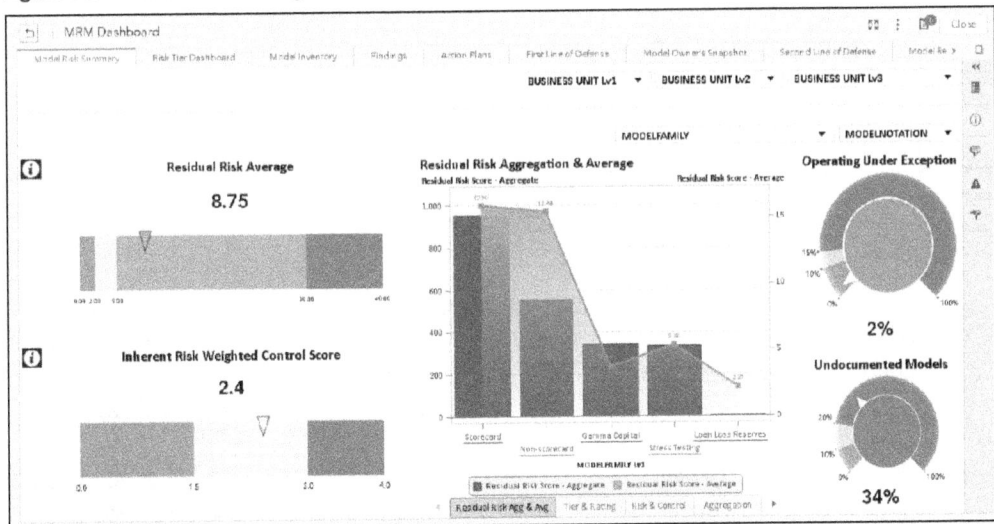

Importing Reports Using the SAS Add-In for Microsoft Office

After you have viewed or enhanced your dashboard reports, you can import them to Microsoft Office using the SAS Add-In for Microsoft Office. This client application enables you to generate reports and refresh them directly from Microsoft Excel, Word, PowerPoint, or Outlook.

Example: Importing the Dashboard Report to Microsoft PowerPoint

Suppose you want to create a PowerPoint slide deck to import and display the entire MRM dashboard. Do the following:

1. Open Microsoft PowerPoint.
2. It is recommended that you use a blank layout for any reports you import. Click the **Home** tab on the ribbon, and select **Layout ▶ Blank**.
3. Click the **SAS** tab on the ribbon, and select **Reports**.
4. Navigate to the folder where your reports are stored (in this example, **SAS Folders/Products/SAS Model Risk Management/Reports/**).
5. Double-click **MRM Dashboard**. The SAS Add-In for Microsoft Office window appears, displaying the dashboard that you selected.

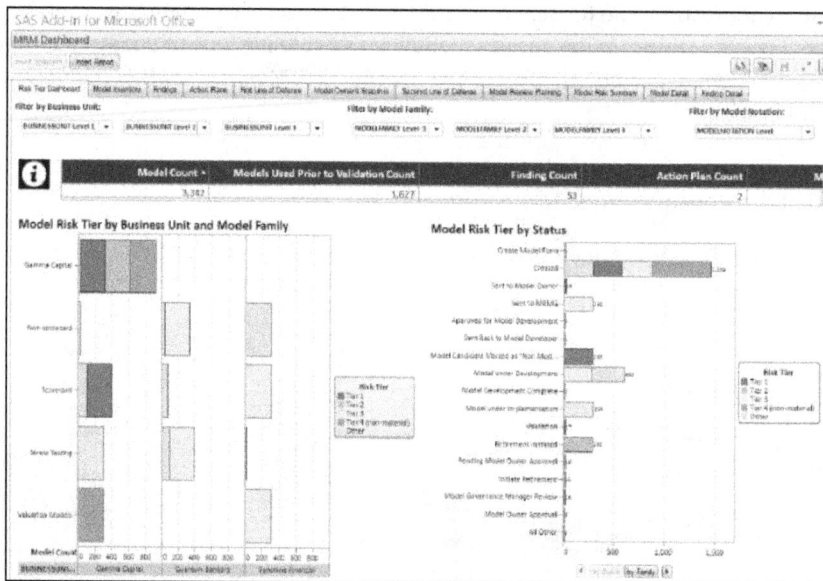

6. Resize the SAS Add-In for Microsoft Office window as needed to view and modify the report. For example, you can apply filters to the report using this window. (**Note:** Filter selections and filter boxes are not inserted into the slide deck, so you should apply these filters before inserting the report.)
7. Click **Insert Report** to begin inserting the dashboard into the PowerPoint deck. During this time, you might be prompted to apply settings to any tables you are adding to the deck.

The dashboard is not automatically formatted for the target Microsoft Office product. After you have imported the dashboard, you might need to perform some maintenance on the appearance of each report object in the slide deck, spreadsheet, or page. In addition, some report objects, such as word clouds, are better presented in the SAS Visual Analytics Viewer, where more details can be seen. You can also choose to import individual report objects.

Demo 7.2: Creating a PowerPoint Slide from the MRM Dashboard

1. Create a connection profile to the metadata server in Microsoft PowerPoint.
   a. Open Microsoft PowerPoint.
   b. Select **Blank Presentation**.
   c. Click the drop-down list for New Slide.
   d. Select the blank slide.
   e. Delete the first slide.
   f. Click the **SAS** tab on the ribbon.
      1) Select **Tools ▶ Connections**.
      2) Click **Add** to add a connection.
      3) For **Connection**, use **sasbap.demo.sas.com**.
      4) For **Description**, the title will auto populate.
      5) For **Port**, use **80**.
      6) For **User**, enter **chunter@saspw**.
      7) For **Password**, enter **Orion123**.

2. Import the dashboard report to Microsoft PowerPoint.
   a. On the SAS ribbon, select **Reports**.
   b. Sign in as **chunter@saspw** using the password **Orion123**.
   c. Navigate **to SAS Folders ▶ Products ▶ SAS Risk Governance Framework ▶ Reports ▶ MRM**.
   d. Double-click **MRM Dashboard**.
      1) Resize the SAS Add-In for Microsoft Office window as needed to view and modify the report. For example, you can apply filters to the report using this window. (Note that filter selections and filter boxes are not inserted into the slide deck, so you should apply these filters before inserting the report.)

      2) Click **Insert Report** [⧉] to begin inserting the dashboard into the PowerPoint deck. During this time, you might be prompted to apply settings to any tables you are adding to the deck.

      3) You might need to adjust some of the pages to look appropriate.

End of Demonstration

Staging SAS Visual Analytics for SAS Model Risk Management

SAS Visual Analytics is an application that uses SAS High-Performance Analytics technologies to visualize, explore, and report on huge volumes of data very quickly. This section demonstrates how to stage your environment so that you can create and save reports and visualize and explore SAS Model Risk Management data.

Before you can begin the steps in Demo 7.3, the administrator must have proper authorization on the operating system and metadata level to start the server. For example, on Windows systems, ensure that the user is a member of the SAS Server Users group. (This group should have the **Log on as a Batch Job** operating system privilege.) The administrator must also be a member of the MRM Data Administrators group in metadata.

Demo 7.3: Viewing the Dashboard in SAS Visual Analytics

1. Start the MRM LASR Server.
   a. Sign in as **Frank** using the password **Orion123**.
   b. Select **Model Risk Management**.
   c. Click the **Show Applications** menu in the upper left corner.
   d. Click **Administrator**. The Visual Analytics Administrator opens.
      1) Click the **LASR Servers** tab.
      2) Select **Risk Gov Frwk LASR Analytic Server**.
      3) Click the **Start** button.
      4) When the status shows that it is running, sign out and close Chrome.
2. Upload the dashboard data.
   a. Open SAS Management Console.
   b. Log on as **SAS Admin**.
   c. Expand **Schedule Manager**.
      1) Right-click **Load_Live Data**.
      2) Click **Schedule Flow**.
      A logon prompt appears.
      3) Enter **Frank** with the password **Orion123**.
      The Schedule Flow should say **Run Now**.
      4) Click **OK**.
      5) Click **OK** when the successfully scheduled window appears.
      6) Right-click **Refresh_Model_Monitoring_Data**.
      7) Click **Schedule Flow**.
      8) Run Now should appear.
      9) Click **OK**.
      10) Click **OK** when the information window opens that states that the flow has been scheduled to run.
   d. Close SAS Management Console.

3. View the dashboard in SAS Visual Analytics.
   a. Sign in as **chunter@saspw** using the password **Orion123**.
   b. Click the **Dashboard** link.

   The dashboard opens.

   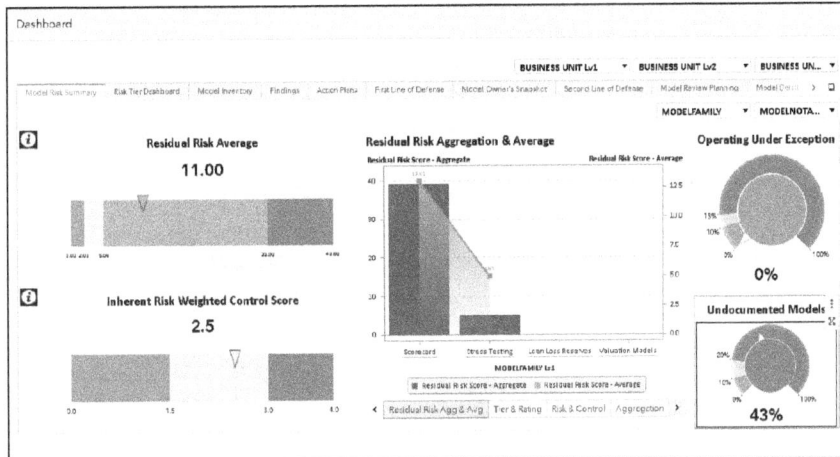

   Across the top is a tabbed navigation bar to other reports.
   c. In the Model Risk Summary, notice that Residual Risk Average is 11.00.
   d. Click ⓘ to see the methodology behind this report.
   e. Explore the reports on this page.
   f. Navigate to the **Model Inventory** tab.
      1) Locate the **Retail Credit EAD** scorecard.
      2) What is the inherent risk rating of this model?
   g. Double-click on the model.
      1) What is the risk tier?
      2) What is the control score?
   h. Click **First Line of Defense**.
      1) In the Model Owners chart, click once on the line for **Carrie Hunter**.
      2) Right-click on this line.
      3) Select **Link navigation**.
      4) Select **MRM Dashboard – Model Inventory**.
      5) Select the first link.
      6) Double-click **Retail Credit EAD Scorecard.**
      7) Under Model Summary, double-click the first instance of **Retail Credit EAD Scorecard**.
      8) Notice that it opens the model in a new tab.
   i. Return to the previous tab.
   j. Sign out.

   End of Demonstration

SAS Model Implementation Platform Integration

SAS Model Implementation Platform integrates with SAS Model Risk Management at different points in the model implementation life cycle, as shown in Figure 7.4.

Figure 7.4: Model Implementation Life Cycle

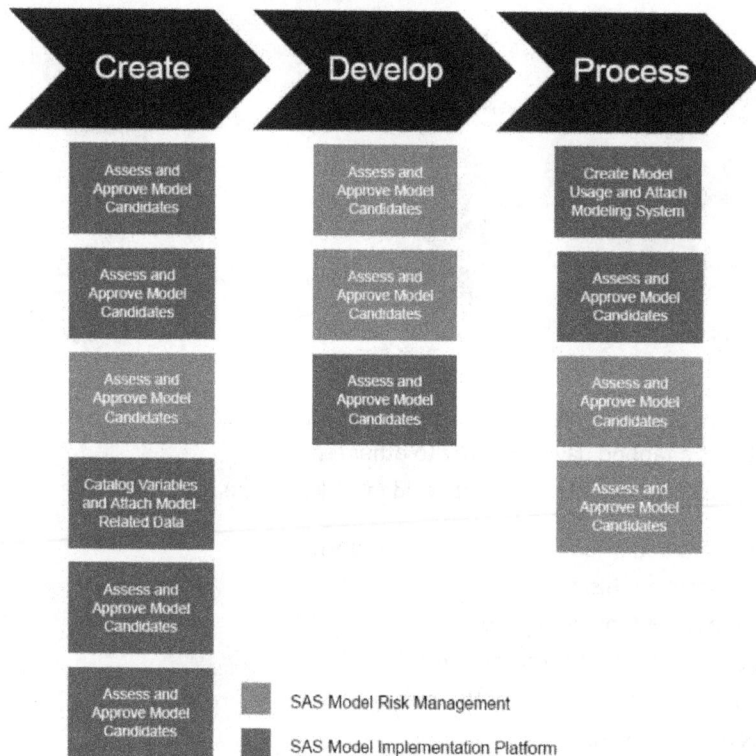

The life cycle in Figure 7.4 assumes a process for developing a new model. For existing or pre-developed models or modeling systems that are being brought into the SAS ecosystem, these model-related objects can be created in Model Implementation Platform first and then imported to the model inventory for review, validation, and approval.

SAS Model Implementation Platform enables financial institutions to build the types of credit modeling systems that are required to satisfy regulatory standards. Model Implementation Platform enables you to implement complex and computationally intensive systems of credit risk models.

Model Implementation Platform enables you to perform the following tasks:

- load individual (sometimes called atomic) models into the system
- create model groups, which coordinate the relationship among atomic models
- override the output of the models that are associated with a model group

Model Risk Management enables you to export models that you have been storing in Model Risk Management and import those models into Model Implementation Platform. These models can be used in creating model groups, developing and debugging user-defined logic, running scenarios and simulations, and so on, with the ultimate goal of creating a modeling system.

Exporting models for use in Model Implementation Platform is done through the Model Inventory page. You can import existing modeling systems from Model Implementation Platform into Model Risk Management. The metadata from this modeling system, including information about models and model groups, is then stored in the Model Risk Management system, and any models not already entered can go through the model inventory workflow. The import of modeling systems is done by attaching these modeling system packages through the Model Usage page.

After the models have gone through validation and the model usage has been reviewed and approved, you can then load modeling systems to the Model Implementation Platform production environment.

Figure 7.5: Export Process

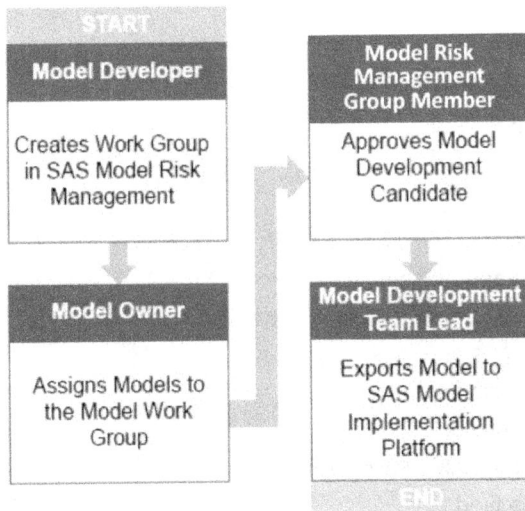

Demo 7.4: Exporting a Model from SAS Model Implementation Platform to SAS Model Risk Management

1. Export models for use in SAS Model Implementation Platform.
    a. Sign in as **Frank** using the password **Orion123**.
    b. Click the **Model Inventory** category button from the navigation bar.
    c. On the Models subcategory tab, in the Model Inventory table, click **New** to create a new model. The Create Model page appears.

d. Expand the **Classification** selector.

1) Click **Manage**.

2) For **Business Unit**, select **Sunshine Financial ▶ Wholesale banking**.

3) For **Model Family**, select **Scorecard ▶ Credit ▶ Retail ▶ EAD**.

4) For **Workgroup**, select **Public**.

5) Click **OK**.

e. Enter the Basic model information.

1) For **Model ID**, enter **M_PD001**.

2) Enter **PD_Model** for **Model Name**.

3) In the **Description** field, enter **This model is to test the export to MIP system.**

4) For **Materiality**, select **Moderate**.

5) For **Criticality**, select **Moderate**.

f. Verify the stakeholders.

g. Verify that the following people are listed in their appropriate group:

1) Model Developer – **Carrie Hunter**

2) Model Owner – **Carrie Hunter**

3) CoE Coordinator – **Paul Melhoff**

4) Model User – **Ann Kellog**

5) MRMG group member – **Kirsten Feldman**

6) OpRisk manager – **Raul Nunzio**

h. Click the **Model Metadata** tab.

1) Select **No** for **Is this a Non-Model?.**

2) Select **Bank Developed** for the **Vended Type** field.

3) Selects **No** for the **Model to be uses before validation?** field.

i. Click **Additional Metadata**.

1) Select **Constant** for the **Model Form** field.

2) Enter **PD** for the **Model Category** field.

3) Select **Probability** for the **Result Variable Form** field.

4) Enter **0.8** for the **Constant Value** field.

j. Click **Save** to save the model.

k. From the **Workflow actions** menu, select **Send to MRMG** to send the model to the MRMG for approval.

l. Sign out.

2. The MRMG member approves the model development candidate.
   a. Sign in as **kfeldman@saspw** using the password **Orion123**.
   b. From the Tasks table, select **PD_Model**.
   c. Click the **Model Metadata** tab.
   d. Enter **04/15/2019** as the date entered on the inventory.
   e. Click **Workflow actions**.
   f. Select **Approve Model**.
   g. Sign out.
3. Validate the model.
   a. Sign in as **Frank** using the password **Orion123**.
   b. Click **Development Details**.
      1) For **Percent Complete**, select **Completed**.
      2) For **Is conceptual soundness documented**, select **Yes**.
      3) For **Is Peer Review Complete**, select **Yes**.
   c. Click **Workflow actions**.
      1) Select **Peer Review #1 Completed**.
      2) Provide a change reason.
   d. Click **Save**.
   e. Click **Workflow actions**.
      1) Select **Development Complete**.
      2) Provide reason text.
      3) Click **Save**.
   f. Click **Workflow actions**.
      1) Select **Peer Review #2 Completed**.
      2) Provide reason text.
      3) Click **Save**.
   g. Click **Workflow actions**.
      1) Select **Peer Reviews Completed**.
      2) Provide reason text.
      3) Click **Save**.
   h. Click **Workflow actions**.
      1) Select **Model Validation Cycle Complete**.
      2) Provide reason text.
      3) Click **Save**.
   i. Click **Workflow actions**.
   j. Select **Model Implementation Complete**.
   k. Click **Close**.
      You should see **Model ID MPD001** and **PD_Model** listed.

4. The model team lead exports the model.

   a. Export the model for implementation. Click in the upper right corner.

   b. Enter an appropriate root folder name such as **PDModels**.

   c. The Export for Implementation dialog box appears, displaying the number of items being exported. The export might take some time, depending on the size of the export. When the export completes, the dialog box closes and the SAS files are created in the SAS Model Implementation file system.

      Note: By default, the number of models you can export at one time is limited by the exportMipSizeLimit parameter specified in the Model Inventory Query Page screen definition.

   d. Sign out.

5. The model developer imports the model.

   a. Using Windows Explorer, navigate to **D:\opt\sasinside\SASConfig\Lev1\AppData\SASRiskWorkGroup\groups\Demo1\SASModelImplementationPlatform\input\models\PDModels**.

   b. Verify that there are models in this directory.

   c. Sign in to Model Risk Management as **chunter@saspw**.

   d. Click the **Show Applications** menu.

   e. Click **SAS Model Implementation Platform**.

   f. If there is an error message, click **OK**.

   g. Click [icon] to display the models page.

   h. Click [icon] to display the Load Models window.

   i. Select **PDModels**.

   j. Click **OK**.

      The model will load.

   k. Sign out.

   End of Demonstration

Appendix A: SAS Model Implementation Platform

Overview

In this appendix, you will learn about the SAS Model Implementation Platform. This product is designed primarily for model developers, model implementation teams, and analysts tasked with generating loss forecasts for stress testing or financial reporting. It is also a central component of the SAS CECL, IFRS 9, and stress testing solutions.

The platform provides a controlled framework for testing and implementing statistical models. Through the automated handling of generalized procedural logic, the platform allows for complex models, such as state transition models, to be implemented quickly and easily. Models are automatically executed in parallel across a cluster of machines for faster performance. Developers can also analyze and compare the results of their models quickly and easily using built-in tools for backtesting, sensitivity analysis, and attribution analysis.

The task of implementing complex systems of credit risk models is something that many financial institutions struggle with. This process involves writing code to combine a portfolio of loans, economic forecasts, and estimated models to officially forecast expected losses, balances, and cash flows over time. There is often a skill mismatch because modelers are more often trained in econometrics or statistics rather than software development.

The SAS Model Implementation Platform streamlines this process for consistent implementations across an organization. SAS provides a built-in framework for rapidly implementing complex systems of models that can be executed efficiently. The results can be dynamically explored to assess each forecast.

Business Use Cases

Some of the common use cases for SAS Model Implementation Platform (MIP) include the following:

- Life of loan forecasting
- IFRS 9
- CECL
- Enterprise stress testing
- Mortgage models

MIP is mostly geared toward any type of loan loss forecasting model. More complex models like mortgage models are especially strong use cases.

Different firms use a wide variety of modeling methodologies, as shown in Figure A.1.

Figure A.1: MIP Methodologies

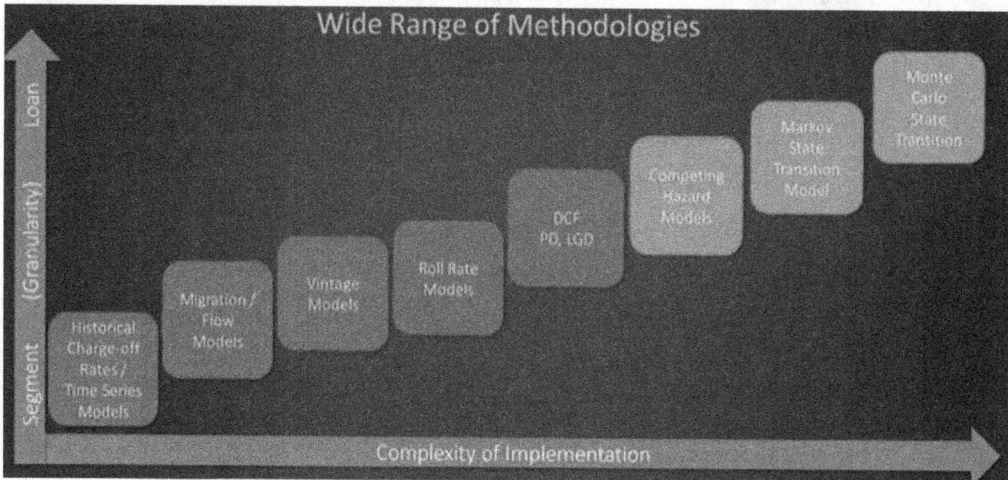

MIP does an excellent job of handling models of many different types. However, it is in the more complex modeling structures, such as the Monte Carlo State Transition framework, where MIP really shines.

Benefits

The real value of MIP lies in streamlining the pipeline between estimating models and implementing them into an integrated system. This is most important when you have many models that need to work together in concert, also known as atomic models. Oftentimes, institutions employ a team of econometricians to build the models and then have another team of developers who code in something like C++ to actually the implement models. MIP makes this pipeline much smaller so that the people who estimate the models can actually implement the models during the estimation process. This allows them to get instant feedback on the results of their model during the development process so that they can make better models and get the results very quickly. Often it can take six to nine months between when sets of models are estimated to when they are implemented to see results. MIP makes that time much, much shorter.

Another thing that MIP does is promote broader model use within an organization. Usually, the people who are building the models own the code and are the ones who are able to actually run the model. So whenever an ad hoc request to get a model result comes from the organization or from regulators, the request has to go through the small set of people who own the code and know the code. With MIP, the models are in a platform where they can be accessed by many different people within the organization. Analysts who don't even know how to write code can access the models and get meaningful and valuable results from them. MIP helps to facilitate collaboration between analysts and model review teams.

MIP encourages the building of more complex architectures. Some of the more complex types of models are difficult to build, and they require a lot of programming. The platform handles a lot of the *process logic*. The benefit of separating out the modeling logic from the process logic is that some institutions start off with up to 10,000 lines of code. By handling the process logic, the platform drastically reduces the amount of code that institutions have to maintain – as much as a 90% code reduction in some instances. This makes it much easier to review and maintain the code over time.

The next benefit of MIP is that it can significantly increase the productivity of the modeling team. When modelers have to do things like backtesting, sensitivity analysis, attribution analysis, and other ad hoc tasks related to model building, the platform allows those things to be done much more efficiently and in a much more structural way with out-of-the box capabilities. This allows the modelers to more closely focus on building better models and makes ancillary things easier for them.

Related to the much-reduced codebase is the reduction of "key person" risk. When you have 10,000 lines of code that only a few people in the organization can understand, and one of those people leaves the organization, then the institution is in a lot of trouble because it doesn't have people who can maintain the code going forward. Because the amount of code that has to be maintained in MIP is much smaller in a much more controlled environment, there is much less risk of an institution being in trouble when a key person leaves.

Finally, the most important benefit of MIP for many banks is that, whether you are doing simple or complex models, there is a desire to have a transparent, controlled, and auditable process.

Core Components

Now we will cover the following core components of the Model Implementation Platform:

- **User Interface:** quickly load estimated models

- **Atomic Model Inventory:** an inventory of individual atomic models

- **Model Groups:** combine individual atomic models that execute together and generate desired outputs

- **High-performance Risk Explorer:** explore model results across any number of scenarios and cross-classifications in real time via an in-memory interface

- **Backtesting:** easily assess model accuracy by comparing model results to actuals

- **Model Sensitivity Analysis:** enables users to shock economic risk factors through an intuitive user interface

- **Out-of-the-box Attribution Analysis**: compare the results between two runs via iterative swapping of risk factors

User Interface

The user interface allows many different people within the organization to have access to the models, see them, and be able to understand what's going on in the model. There are two different types of key users who use this interface. One group of users is the model developers or model implementation team who set up the models and get them into the platform along with details about the models. The second set of users are the people who want to be able to execute or run the models. Once the models are in the platform, some users load the portfolio, load the economic data, and execute those models to get the results. Once the results are in the platform, then those results can be explored.

Atomic Model Inventory and Model Groups

Within the setting up of the models there is the Atomic Model Inventory. This is where you load the estimated model or single result of an estimation procedure. These single models are loaded into the platform. Then they are collected in model groups. Model groups are one or more atomic models that are put together in a meaningful way to give a specific desired result – for example, lifetime losses, monthly losses over time, or default dollars. This is where the implementation process really occurs.

High-performance Risk Explorer

Once a model has been executed through the user interface, you can dynamically explore your results in the high-performance risk explorer. These results are dynamically explored in-memory. You can drill up or drill down. You can see results across many different dimensions, also known as cross-class variables. This is very useful if you are trying to drill down to find potential problems in a model run across your portfolio. Or, if your model is not performing on a particular segment, you can really drill down as far as you want to go, even all the way down to the low level to find potential problems in your models.

As mentioned earlier, modelers are not the only ones who can do this. Any analyst who has access to the platform can do this sort of work. It extends the number of people who can be involved in the interpretation of your model results and driving that value throughout the organization.

Backtesting

Backtesting can really help you in getting additional results around the modeling process. Backtesting often has to be done on a monthly or quarterly basis to make sure that your models are performing as intended. MIP makes it very easy to run a model at some point in the past, for example, one, two, or three years back in time. You run it forward and overlay the actual results of the output variables that the model is creating; that is, you can overlay your actuals with your predicted values. Similar to the way you can dynamically explore the results of a forecast, you can also dynamically explore the results of your backtesting. This is very important for being able to work with regulators and validation teams.

Model Sensitivity Analysis

It is very important to get instant feedback, particularly if you have a big system of models. Let's say you have 30 atomic models that are working together within the system of models. You need to know that the output of this model group is sensitive in appropriate ways to important factors. What you get out of the box with MIP is the ability to choose various shocks – for example, up and down 10% or 20% of these economic factors. With a few clicks, you can set up and execute these shocks. Then you are able to explore the results of the shocks just like you can explore the results of any of your model runs. This enables you to quickly respond to questions from anyone who asks how sensitive your models are to important factors.

Out-of-the-box Attribution Analysis

Attribution analysis enables you to take a run from a past quarter or year ago and compare it to results run today. The first question anyone will want to know is: what changed between those two runs? A host of things could have changed, but you want to identify which factors have changed. For example, there could be new loans at a deep portfolio, some loans might no longer be in your portfolio, or economic factors may have changed. Loans might also have aged, so even if the same loans are still in the portfolio, they are different from what they were at some point in the past. The models themselves may also have been re-estimated or changed. Maybe tuning factors have been changed in the model. So how do you determine which of these many things might have changed? MIP enables you to set up the analysis very conveniently and get the result in a waterfall graph that shows all changes.

In this section, you have been introduced to a summary of some of the key things that you are able to do within the Model Implementation Platform. Next, we will walk through a demo of the platform.

Demo

In this demonstration, we will do two things. First, we will walk through the process of how you would implement a model into the SAS Model Implementation Platform. Once that model is implemented, the second demonstration shows how to create forecasts and learn about the different types of analysis you can conduct.

Implement a Model

The SAS Model Implementation Platform (MIP) is an implementation product. The basic setup of actually creating a model can be broken down into two phases. The first phase is *estimation*, where there is a set of historical data and some set of models is estimated depending on the framework to create the basic model equations. The MIP picks up from this point. Once the models are already estimated, the models need to be combined to produce a forecast given a new portfolio.

Let's look at an example of how to implement estimated models in the system. We will use a current to d30 model that has already been created. Open the SAS Model Implementation Platform. Open the **Risk Model Editor** by clicking on the icon in the upper right corner as shown in Figure A.2.

Figure A.2: Open Model Risk Editor

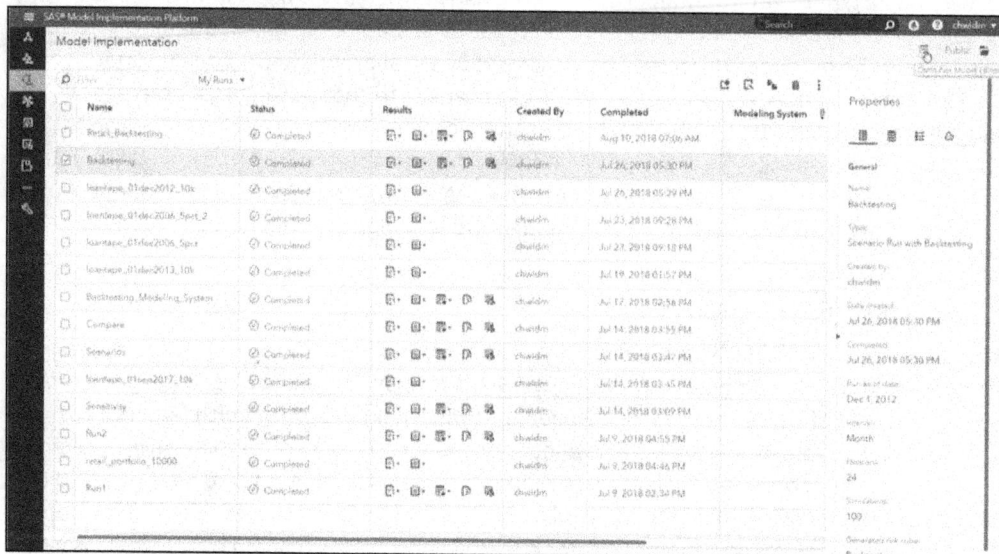

After the Model Risk Editor opens, double-click the model that you want to implement from the list of models shown. In the **Model Properties** tab, enter some metadata about the model including giving it a Model ID, a Model name, saying who developed it, entering a model description, and adding a documentation link.

Figure A.3: SAS Risk Model Editor

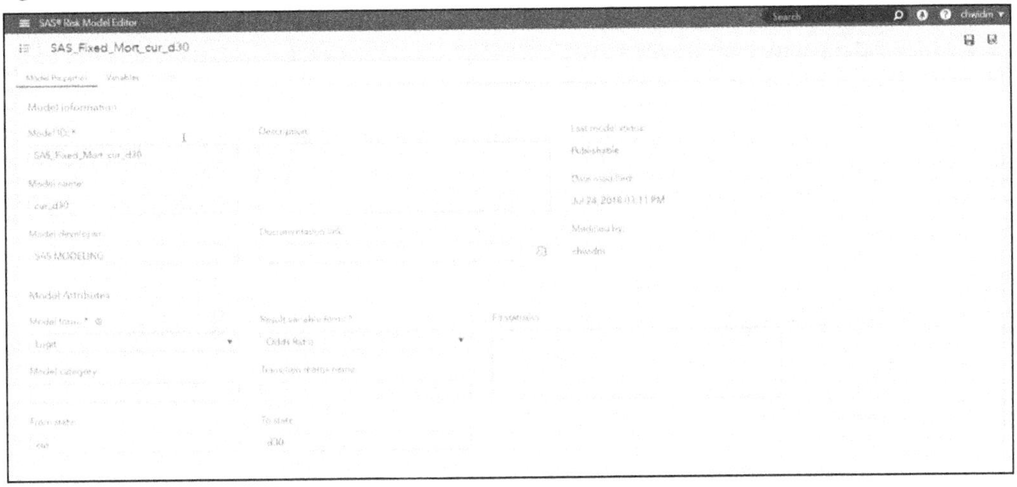

You can choose a **Model form** from the drop-down menu, which are the prebuilt forms that MIP supports. If none of the options fit what you want, you can select **Custom** or choose **Code** and write the code yourself. Use the **Results variable form** drop-down menu to tell the system what type of form the result will produce. If the model is a transition model like the one shown in Figure A.3, you would enter the **From state** and the **To state**. Once we get to the model group, we will see how that will be used.

If the model form contains variables – most of them do, but the constant model does not have variables, for example – then you will see the **Variables** tab. Click on the tab. To create the variables, click the ⏎ icon to go to load a coefficient file. This allows you to select locally a coefficient data set that can either be a SAS data set or a CSV file. Once you load the file, you will get an entire listing of variables with the variable name and coefficients, as shown in Figure A.4.

Figure A.4: Variables Tab

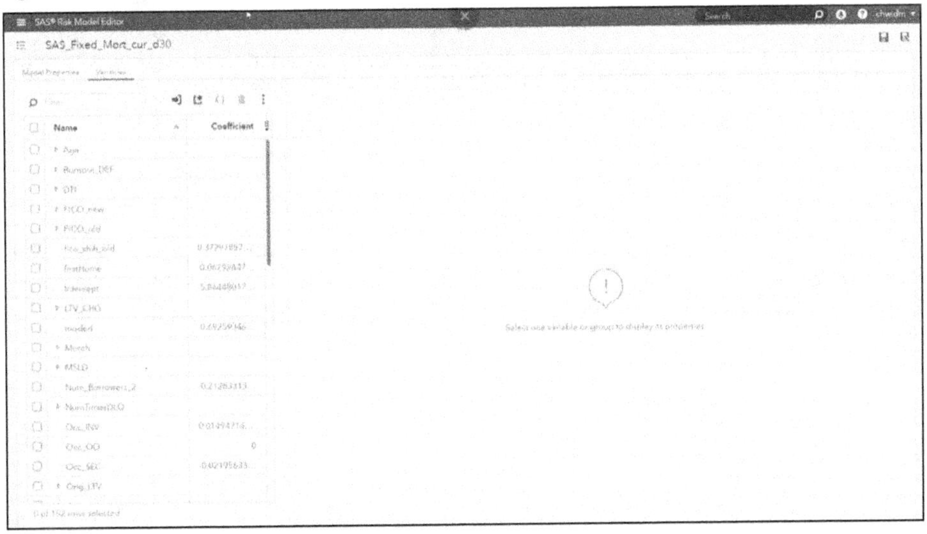

You can see in Figure A.4 that some of the variables are grouped together with no coefficients. That's because these are spline variables that the system has automatically grouped together. Underneath each group, you can see the individual knots of the spline. Click on any of the variables in the left pane to look at the associated metadata.

When you are building a new variable, you need to enter in the group name, calculation type, and a little bit of metadata about each one to tell the system what this variable is and explain to it how it will work. With the spline variable in Figure A.5, you can see the individual knots in the group. Click on the **Show graph** icon to see its effect.

Figure A.5: Show Graph

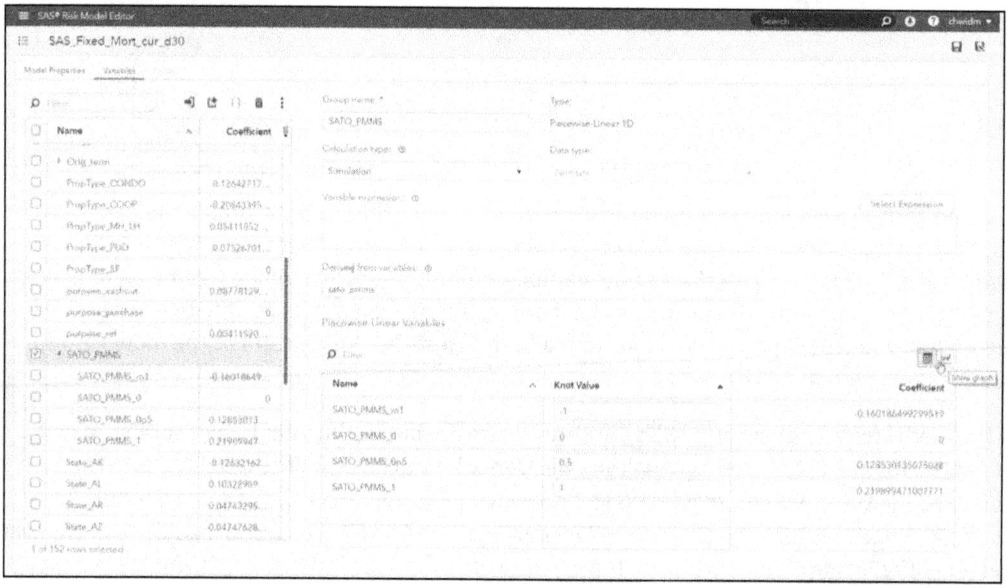

Click on the icon in the upper right corner to return to the list of models. For every model that you want to use, you need to enter in some metadata about the model and each of the variables. For example, in a simple hazard framework, you might have a default model, a prepay model, and a severity model. After you build all three models, you would enter them into the system, load the variables, and define the variables. It will take a few minutes for each of them.

Publish a Model

Once you have defined all variables for model, in the SAS® Risk Model Editor window, select the model that you have finished defining by clicking the check box to the left of the Model ID. In the upper right corner, click the **More options** icon, and then choose **Publish Models** from the drop-down menu, as shown in Figure A.6.

Figure A.6: Publish Models

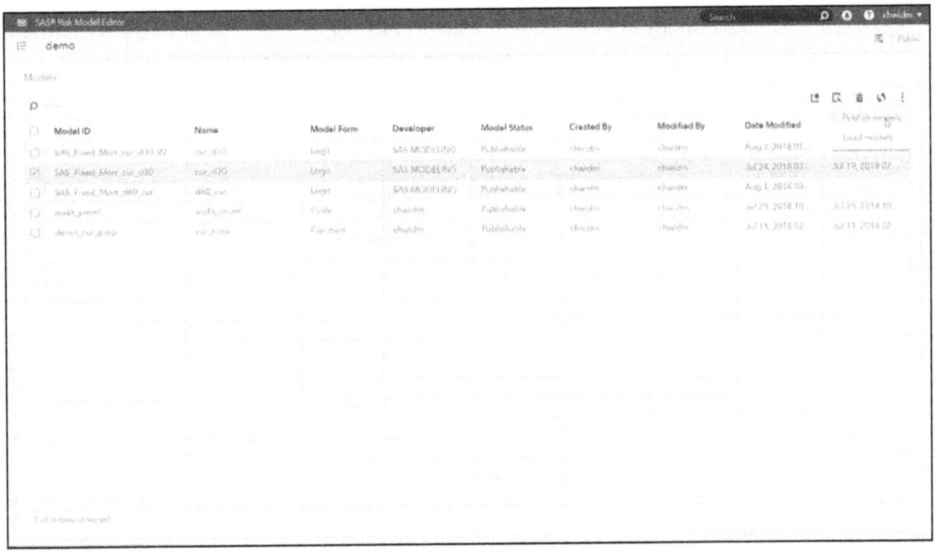

Clicking **Publish Models** sends the model into the Model Inventory. Let's take a look now at the Model Inventory. From the main screen of the Model Implementation Platform, click on the single triangle icon at the top of the left sidebar. You will see that the Model Inventory has a listing of all the models that have been published and are available. (See Figure A.7.)

Figure A.7: Model Inventory

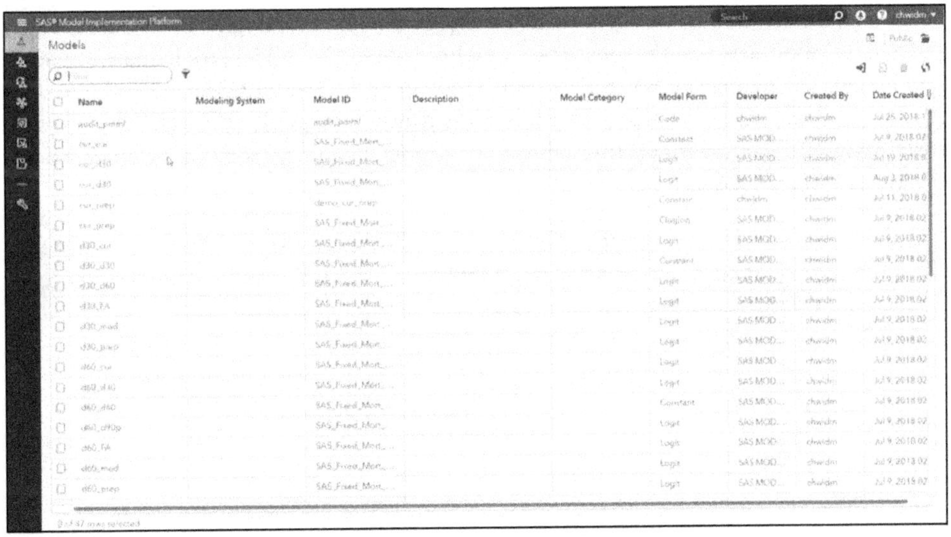

Click on a model **Name** to see the metadata information stored in the database. You can also see the variables along with their individual information. A model in the Model Inventory is ready to use.

Create a Model Group

Now that you have entered individual models, we want to put them together and use them in a system called a *model group*. Click on the second icon on the left-hand sidebar that shows two interlocking triangles. This shows a listing of your model groups. (See Figure A.8.)

Figure A.8: Model Groups

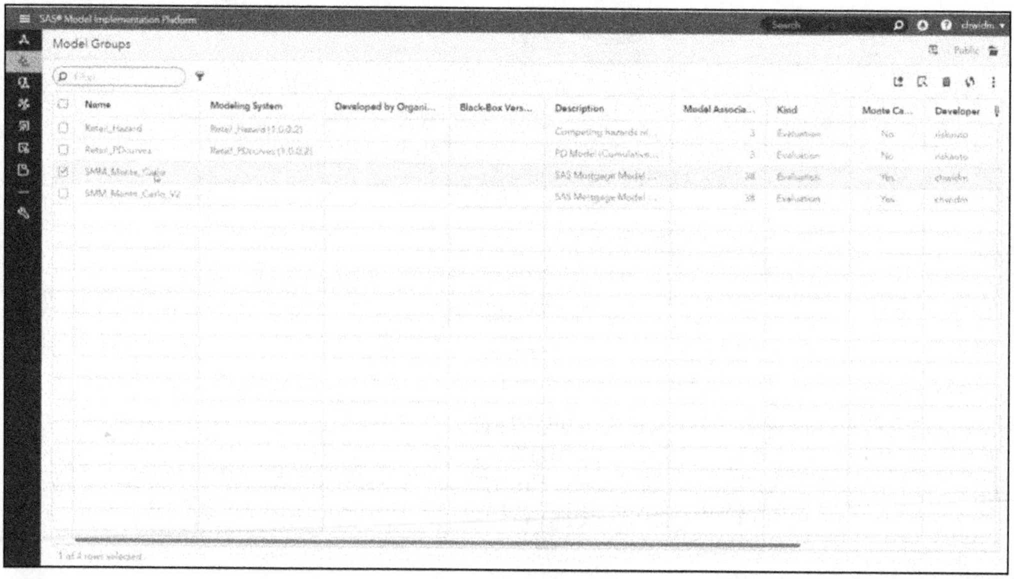

Double-click on a model Name to see a preexisting model group. You can give the model group a name and a description. The system keeps track of who created the model group and when it was last modified. The models in the model group show up in the **Model Associations** table on the left. To add a model, click the **plus** (+) button in the **Model Associations** panel. This opens the **Associate Models** window that is the same as the Model Inventory. Select whichever models you want to add, then click **OK.**

To use these models together to produce a forecast, look at the **User-Defined Logic** pane. (See Figure A.9.) The user-defined logic follows a very simple pattern: derive some variables, score some models, and produce some outputs.

Figure A.9: Model Group

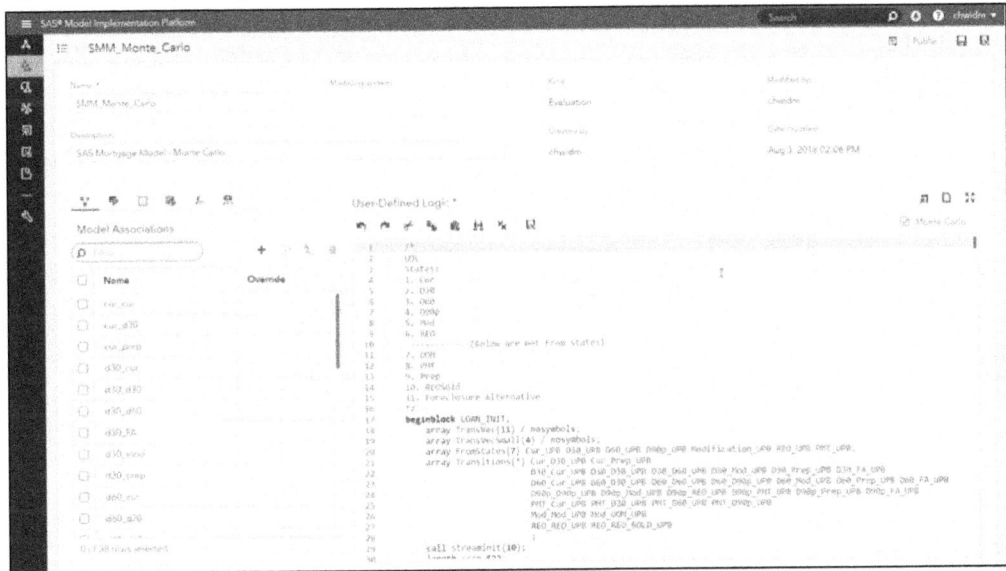

We won't be diving further into the code in this brief demonstration, but if we did, you would notice that it's a relatively small amount of code compared to the complexity of the system that it is building. The reason for that is a lot of the infrastructure tasks such as looping over horizons, looping over simulations, looping over scenarios, parallelizing the logic, handling the output variables, and merging of economic data are automatically handled by the system. The only thing developers need to do is tell the system how the specific model works, what variables the model needs, when to score the models, and what outputs to produce.

Once you have imported your models, written the code that says what to produce, and tell the system what output variables to keep track of, then you are finished. You are now ready to produce a forecast.

Create a Model Group Map

However, there is one final concept to cover before creating a run – creating a model group map. A *model group map* is a traffic cop that picks up loans off of the portfolio and reads a field and tells it which model group to execute. Click the wrench icon in the left-hand sidebar to view the **Configuration** window. Select **Model Group Maps** to create a view model group maps.

We have walked through the basic steps of how a developer might implement a model in the SAS Model Implementation Platform. Once the individual models have been added, the model group has been created, and a model group map has been defined that points to the model groups, we are ready to use them. Click on the **Model Implementation** icon in the left-hand sidebar. This opens the Model Implementation window. Before you can create any forecasts, you need to send the portfolio to the grid.

Click the ⬑ icon in the upper right corner. Select **New portfolio cube run** from the drop-down menu. This opens the **Cube configuration** window. (See Figure A.10.) Give it a name. In the **Portfolio data** field, click the **Select** button. Select the data set of the portfolio that you want to use. Next, tell the system which date is the **Run as-of date**. In the **Risk Grid Configuration** section, tell the system how many nodes and how many threads to run on. By default, you can use the entire grid, or you can use some smaller subset. Click **Submit Run** in the upper right-hand corner.

Figure A.10: Cube Configuration

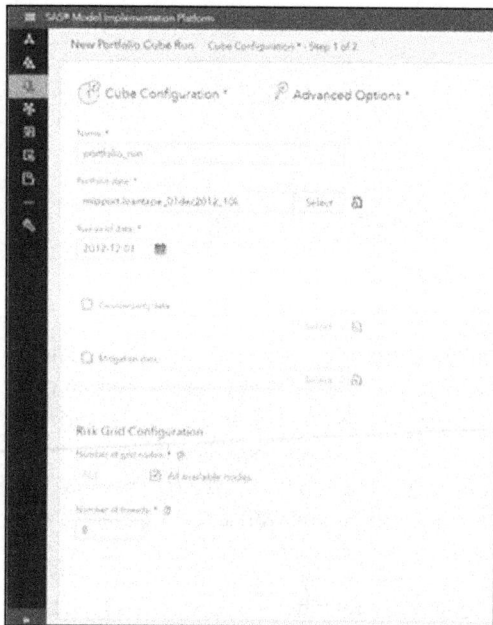

This picks up the portfolio, distributes it out to the grid, and makes the data available for use. The running of the model is separate from the distribution of the data. You can push the portfolio out to the grid once and use it as many times as you would like.

Run a Scenario

Now let's examine a run. Return to the Model Implementation window. The first thing we want to do is create a basic scenario run using a portfolio that has been pushed to the grid, uses the model group map defined earlier, and uses some economic scenarios to create a forecast. Click the ⬑ icon in the upper right-hand corner. From the drop-down menu, select **New scenario run**. This opens the **Edit Scenario Run** window. The first thing you do in this window is select one of the portfolio cubes that have been pushed out to the grid or a previous scenario run.

Next, click on **Scenario Data** at the top of the screen. This is where you will select your economic data. Under **Scenario economic data**, click **Select** and choose the economic data in the file system. If you would like, you can quickly graph the economic data by clicking the graph icon.

Click **Maps and Methods** in the Edit Scenario Run window. Here you choose the model group map to run through. Remember, the model group map points to the model groups that it will use. Under **Evaluation model group map,** click **Select** and choose your model group map. If it is a Monte Carlo model, you will tell it the number of **Simulations** to run through.

There are a few other options that can be selected in this step, and you can use **Advanced Options** and **Execution Options** to customize other aspects of the data and execution. At this point, however, the scenario is ready to be run. Click **Submit Run** in the upper right-hand corner. This goes out to the grid, takes the portfolio cube that you had distributed, runs these scenarios using the model group that you created, and produces the forecast according to the given parameters.

The results are stored in two different places. The individual results are stored on the grid and can be looked at using the risk explorer. Also, depending on the query you look at, the query level determines what results are brought back locally. To launch the risk explorer, click the **Explore results** icon 🗖 in the **Results** column of the Model Implementation window. This opens a **Risk Exploration** window. Here you can interactively evaluate the forecast and compare the results.

In the **Data** panel of the risk explorer, click on **Scenarios** to expand it. You can see all of the scenarios that you ran. Right-click the Scenarios folder and select **Add "Scenarios" to "Crosstab 1"** to add the scenarios to the cross-tab. Next, click **Output Variables** to expand it. Right-click on a variable to add it to the cross-tab. Add a cross-class variable to drill down. Click **Class Variables** and follow the same process to add any class variables to the cross-tab.

Now you can look at the individual results. Right-click inside the cross-tab and choose the first option in the drop-down menu to compare variables. Results can be compared by **Subportfolio, Horizon,** or **All Scenarios by Horizon.** Choose **All Scenarios by Horizon** to compare variables by all horizons and all scenarios. (See Figure A.11 for an example that shows a total portfolio forecast for all losses for every scenario.)

Figure A.11: Compare All Scenarios by Horizon

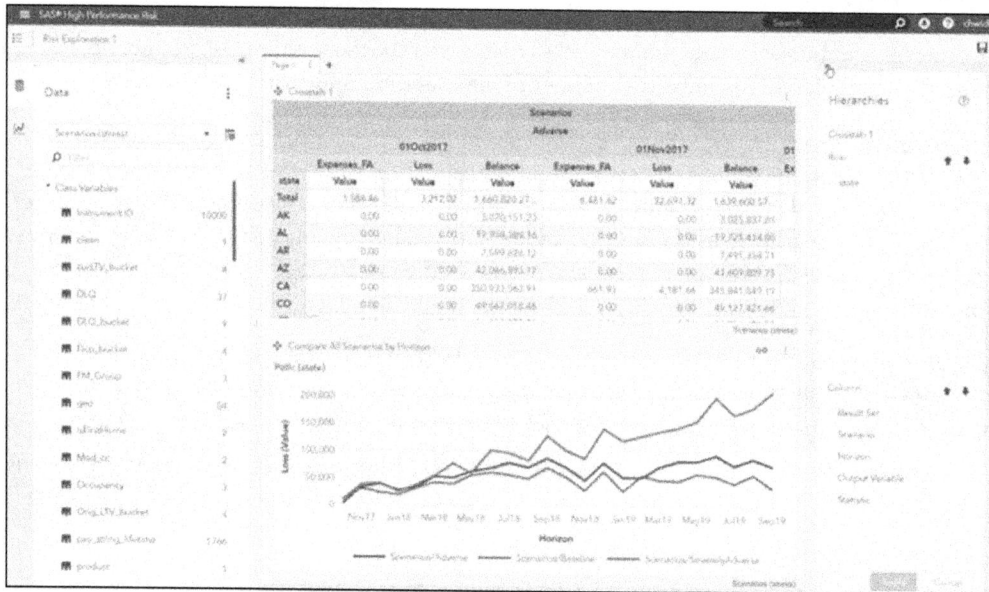

You can adjust the graph by double-clicking on any cell within the cross-tab. With just a few clicks, you can see any of the cross-class variables that you have made available to the system. You can drill down on your output variables by any of the cross-class variables. You can also drill down as finely as you would like to evaluate your results.

Run Other Analyses

Using the model that you have implemented with your model group map to produce a scenario forecast is the most basic type of run and the most common. The goal of the SAS Model Implementation Platform is to make it as quick and easy as possible to run a whole host of different types of analyses. Let's examine a few of the others.

Scenario Run with Backtesting

The first analysis we will look at is a scenario run with backtesting. With backtesting, you have a historical portfolio and you want to compare your model forecast against what happened. This is a common situation that a modeler or an analyst will have to do to demonstrate the accuracy of their model. The SAS MIP makes backtesting very quick and easy by providing it as a built-in run type.

Return to the **Model Implementation** window. Click the ⬚ icon in the upper right corner to create a new run. From the drop-down menu, select **New scenario run with backtesting.** The setup for this scenario looks very similar to the setup of the normal scenario run. Select a **Source risk cube**, select some **Scenario data**, and select which model group map you want to backtest. Here you will get an extra step in the wizard for the backtesting setup. In the **Backtesting** step, you will select an external data set that contains information about what actually happened for the output variables that you are concerned about. In the **Actuals data set for backtesting**, select the data set you want to use. Click the box next to **Calculate variables for backtesting metrics** to select any of the output variables contained in the run and get built-in metrics for those variables. The possible metrics are cumulate percentage error, mean absolute error, mean absolute percentage error, mean error, and root mean square. If you need an additional metric, you can also create your own.

Once you have gone through all of the steps in the wizard, you are ready to run. Click **Submit Run**. Once the run has finished, you can view the results in the Risk Explorer. Once you have launched the Risk Explorer, you will notice in the **Data** pane on the left side under **Scenarios** that you have two scenarios. The first is automatically labeled as **actual**. Those are the results from the actual data set that you loaded. The second scenario is the name of your run. Right-click **Scenarios** and choose **Add all items from the folder "Scenarios" to "Crosstab 1"** to add all scenarios to the cross-tab. (See Figure A.12.)

Figure A.12: Add Scenarios to Cross-tab

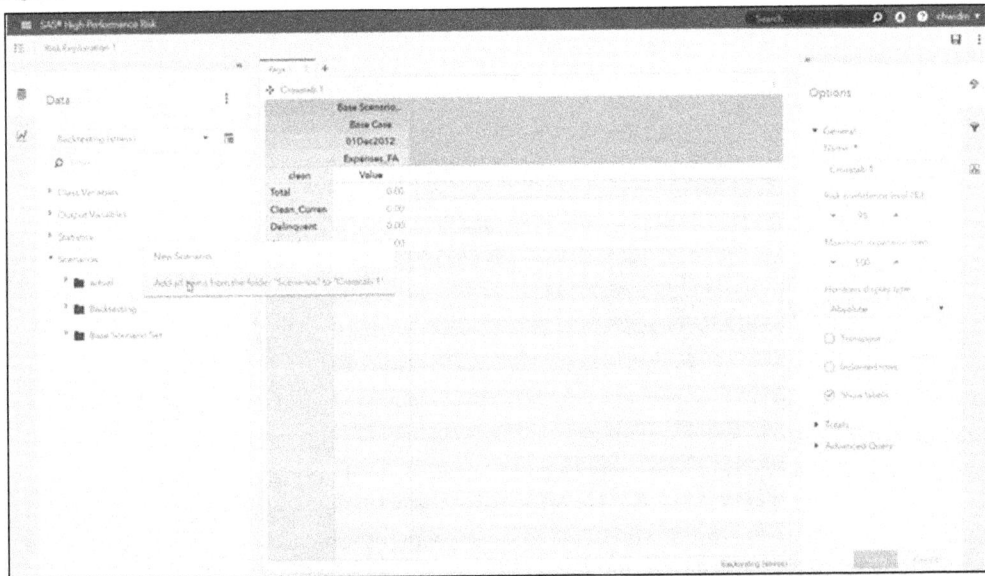

Next, choose the output variables that you are interested in. Click **Output Variables** in the left pane and use the filter search to find and select variables. Right-click on the variable name and add it to the cross-tab. Once you have added a variable to the cross-tab, right-click inside the cross-tab and use the options in the drop-down menu to compare the variable by all scenarios and horizons. This enables you to very quickly plot the results of your forecast against what actually happened. (See Figure A.13.)

Figure A.13: Compare All Scenarios by Horizon

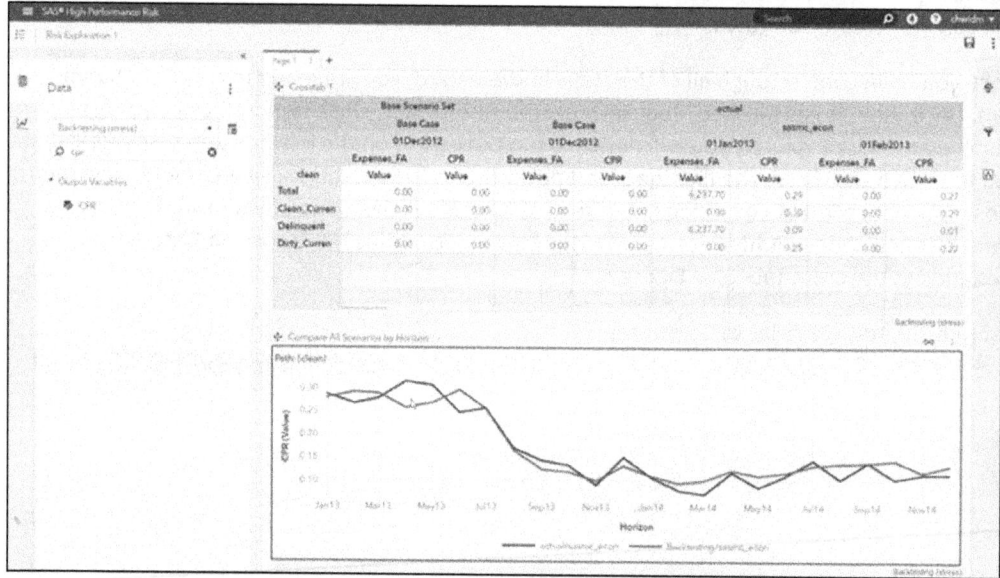

You can drill down and look at any lower cross-class segment. Double-click on any cell in the cross-tab to look at a lower segment.

Sensitivity Analysis

The next type of run that we will walk through is a sensitivity analysis. This is a common type of analysis that a modeler or an analyst might need to perform to demonstrate the sensitivity of their model by shocking som economic factors or numerical portfolio factors. For example, you might want to demonstrate the sensitivity of your prepayment forecasts to interest rates or the sensitivity of your default forecast to unemployment rates.

From the Model Implementation window, click the 🖰 icon in the upper right corner to create a new run. From the drop-down menu, select **New scenario run with model sensitivity analysis**. Again, the setup wizard is very similar to the previous runs that we have set up. You will select a source risk cube and select your scenario data. There is a new step here to create a Model Sensitivity Analysis page. In this step, you can create shocks.

Under **Risk source type**, select which type of factor you want to shock. Under **Variable**, you will select a risk factor group from your economic data. Next, select if you want to create absolute or relative shocks under **Shock type**. In the **Shock values** box, enter the values that you want in a list separated by returns. You can give it a **Floor** and a **Cap** to say that you don't want values to go above or below a certain number. You can also **Fade in** or **Fade out** the shock over a period of months. (See Figure A.14.) When you are happy with what you have entered, click **Apply**.

Figure A.14: Model Sensitivity Analysis Setup

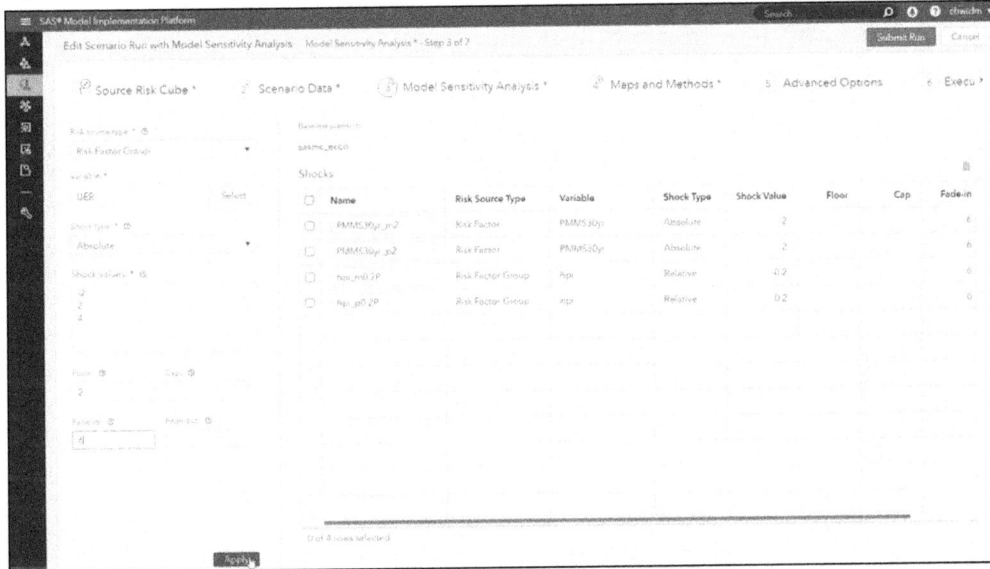

Now you will have created new scenarios that correspond to the shock values you entered. The rest of the setup in the wizard is the same as shown previously. When you have finished configuring your options, select **Submit Run**. The system will run all the scenarios that were created in the sensitivity analysis step.

Go the Risk Explorer to evaluate the results. In the **Data** pane, click **Scenarios ▶ Sensitivity**. You can see the shocks that you created. To evaluate a shock, right-click the shock that you want to see and add it to the cross-tab. You can add as many shocks as you would like. Next, add any output variables that you want to assess by clicking **Output Variables** and then right-clicking the name of the variable to add it to the cross-tab.

Now you can plot the forecast. Right-click inside the cross-tab on the variable that you want to compare. Select **Compare (Variable) by ▶ All Scenarios by Horizon**. Hopefully, the plot will quickly show that your model is very sensitive to shocks. (For example, see Figure A.15.)

Figure A.15: Compare All Scenarios by Horizon

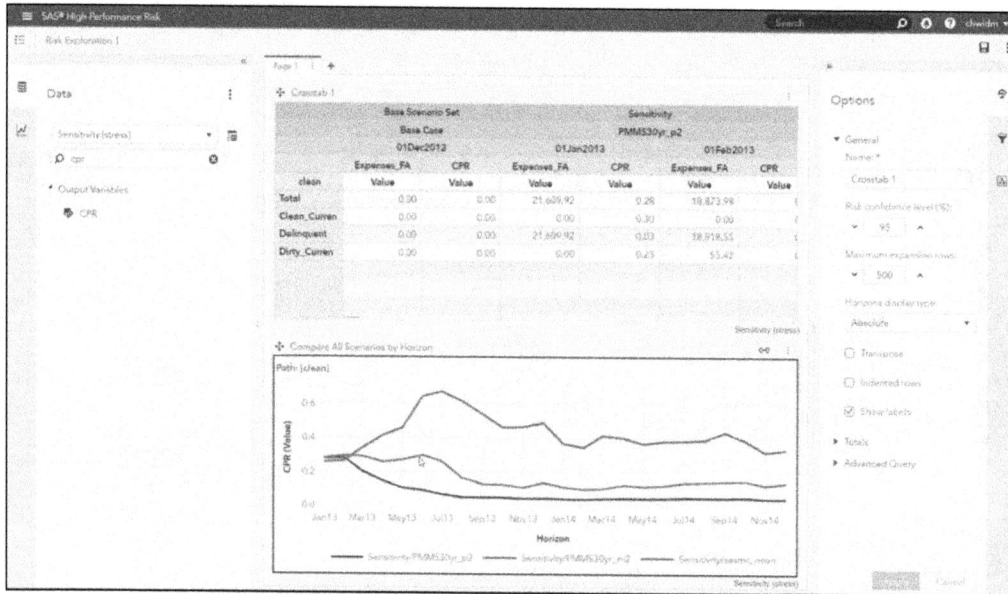

Attribution Analysis

The last type of analysis that will we run is an attribution analysis. A common situation in any sort of regular reporting, for example, stress testing or reserving, is that you will produce a number at one time and then the next quarter you will forecast a new number. And, of course, the questions from senior management are. "Why is there a change? What happened? Can you explain to me why the two numbers are different?" Typically, the process for attributing the differences involves starting at the first run, then iteratively swapping in the factors from the second run to walk forward from the first run to the second run, and explaining why the two numbers differed.

In the most recent release of SAS Model Implementation Platform, the attribution analysis process is built-in for you. In the sidebar on the left side of the Model Implementation screen, there is an icon for Attribution Analysis. Click on the icon of the page with a magnifying glass. This takes you to the **Attribution Analysis** window where you can create a new attribution analysis.

Click the icon in the upper right corner. This opens the **New Attribution Analysis** screen. First, give the analysis a name. Then, in the **Run 1** box, select the first run you want to compare. In the **Run 2** box, select the second run. Click **Apply**. The system now brings in the economic factors, risk data objects, and computed variables. It lists them all in the attribution factors table, as shown in Figure A.16.

Figure A.16: New Attribution Analysis

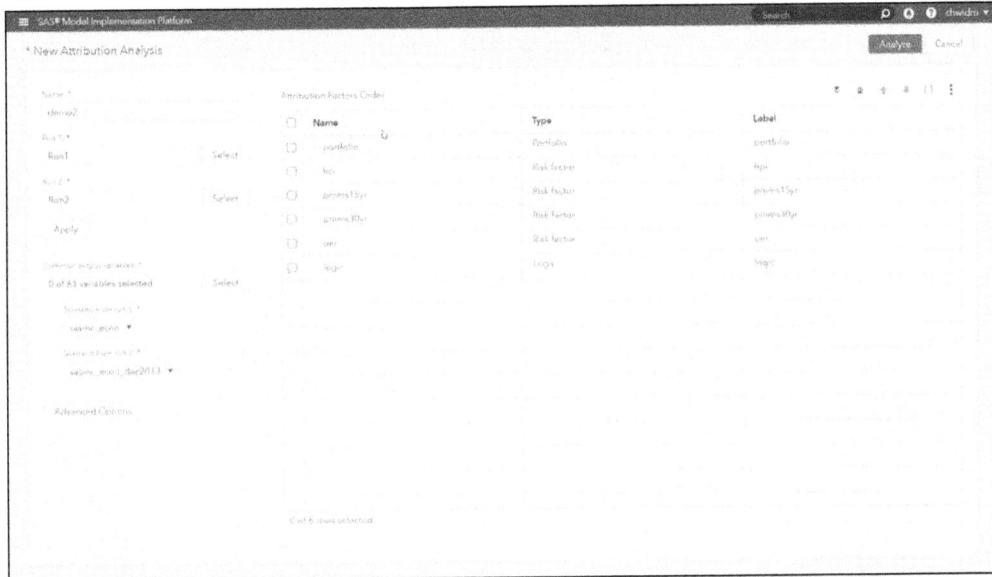

The attribution factors listed in the table are acting as the sequence of updates that the system will go through. It starts at Run 1 and swap in the factors from Run 2 in the order specified in the table. You can change the order in the table by clicking a factor and using the up and down arrows in the upper right to move the factor up or down in the list. However, the portfolio must be first and the logic must be last. You can also group factors. Select any factors that you want to see a grouped impact, rather than the individual impact. Click the **New Group** icon in the upper right corner. A new row appears titled **Group 1**. You can rename the group if you want.

Next, select any output variables that you are interested in. Click **Select** under **Common output variables** to add variables. For example, if you stress testing, the loss variable would be an important one. You can select any two scenarios with a different. So you could compare, for example, base versus adverse from Run 1 to Run 2. Or you can compare all scenarios that have the same name, for example, base versus base, adverse vs. adverse, severely adverse vs. severely adverse, and so on.

Click **Advanced Options** to customize the advanced options. You can set the number of nodes and threads. By default, the generated risk cube is discarded, but if you want to query them later, you can uncheck that option to keep those cubes.

Now you are ready to click **Analyze**. With just a few small steps to set up, you are ready to create and launch this potentially large analysis. Once the attribution analysis is complete, you can view the results. The results are a chart that walk you from Run 1 and ends at Run 2. (See Figure A.17.) For each of the factors in the table that you had set up, you can see which part of the difference between Run 1 and Run 2 is attributed to each of these factors.

Figure A.17: Attribution Analysis Results

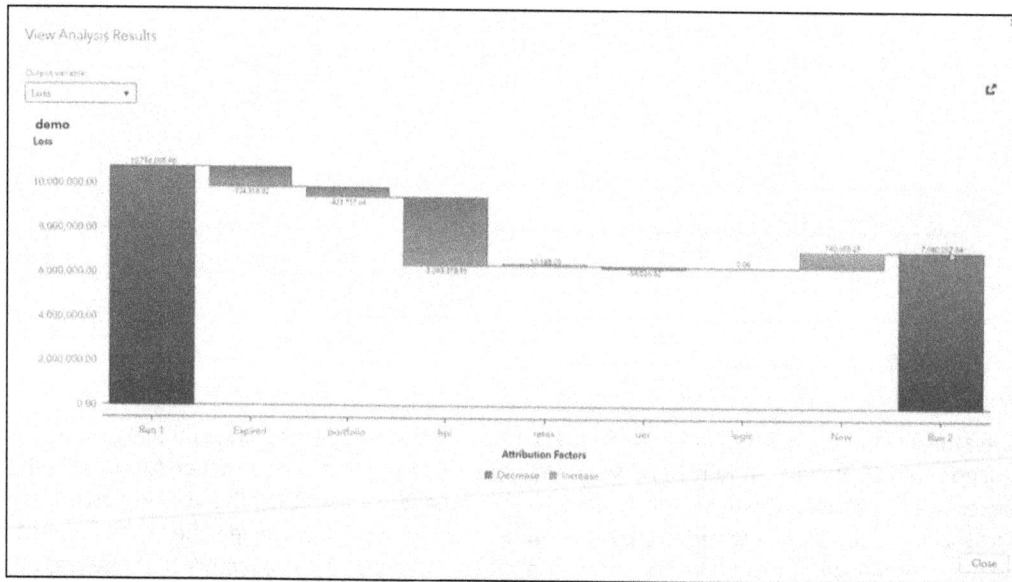

Ready to take your SAS® and JMP® skills up a notch?

Be among the first to know about new books, special events, and exclusive discounts.
support.sas.com/newbooks

Share your expertise. Write a book with SAS.
support.sas.com/publish

sas.com/books
for additional books and resources.

§sas.
THE POWER TO KNOW®